# THE ENGINEERS

# THE ENGINEERS

This catalogue has been published to coincide with an exhibition of books, prints and drawings at the Architectural Association from 4 to 27 March 1982.

The exhibition has been brought together Frank Newby, Senior Partner of F.J. Samuely & Partners, and Julia Elton, Director of B. Weinreb Architectural Books Ltd.

The exhibition and catalogue have been designed by Ron Herron and Tony Meadows.

# CONTENTS





ISBN 0 904503 16 X

# INTRODUCTION



'The basis of engineering is knowledge of the materials being used: knowledge of what they are made of, how they are made, how they are shaped, how they stand up to stress, how they break, how they catch fire, how they react to all the agencies of ruin which are perpetually nibbling at them, how in due course they fall down. This is the true basis of engineering'. So said A.J. Harris in an address to the RIBA in 1960 on 'Architectural Misconceptions of Engineering'. He also stated that engineering was an art and not a science, echoing the remarks made by Torroja in his book *Philosophy of Structures:* 'The process of visualising or conceiving a structure is an art. Basically it is motivated by an inner experience, by an intuition. It is never the result of mere deductive logical reasoning...The achievement of the final solution is largely a matter of habit, intuition, imagination, common sense, and personal attitude. Only the accumulation of experience can shorten the necessary labor or trial and error involved in the selection of one among the different possible alternatives.'

Although these statements come from distinguished engineers, they are readily comprehensible. Names such as Telford and Brunel are household words but their achievements are understood only in the vaguest terms for engineering is generally dismissed as a science beyond the grasp of most people.

This exhibition is about the art of structures. To understand more easily how they are conceived three questions should be asked: why, where, and what is a structure?

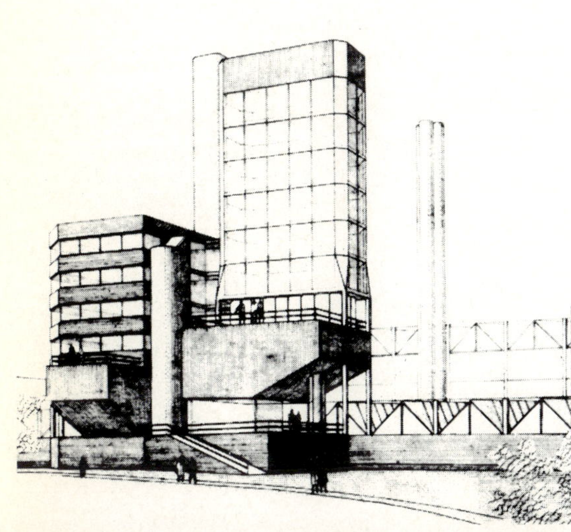

Shelter: Leicester Engineering School, 1959.

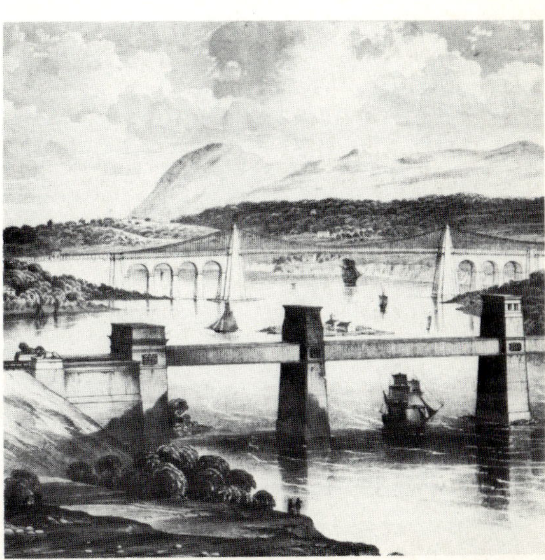

Convenience: The Menai Bridges.

**Why a structure?** Man must build shelters against a hostile environment and as societies grow in size and complexity, so also must the shelters that protect living and working places. Today proposals to cover whole cities hardly seem far-fetched.

Structure and foundation prevent shelters from collapsing, distorting or vibrating, and the engineer has always been responsible for their safety and stability. The architect, on the other hand, is responsible for their design and the interaction of architect and engineer became increasingly significant as the building process moved from a craft to a science-based industry.

Man has also become increasingly mobile and part of the work of the civil engineer is to bridge rivers and valleys with structures of great ingenuity and grace. Indeed the Institution of Civil Engineers defines its profession as 'the art of directing great sources of power in nature for the use and convenience of man.

Advances in the art of engineering have usually come about for specific reasons. The railway era created new problems in scale and loading of its structures as do the offshore drilling rigs today. War, too, poses new problems solved, for example, by inflatable bridges, Mulberry harbours and supersonic aircraft.

**Where a structure?** The site is unique. Weather and ground conditions can vary substantially from one field or continent to another. The availability and quality of labour and materials also vary in time and their use may be impeded by local regulations.

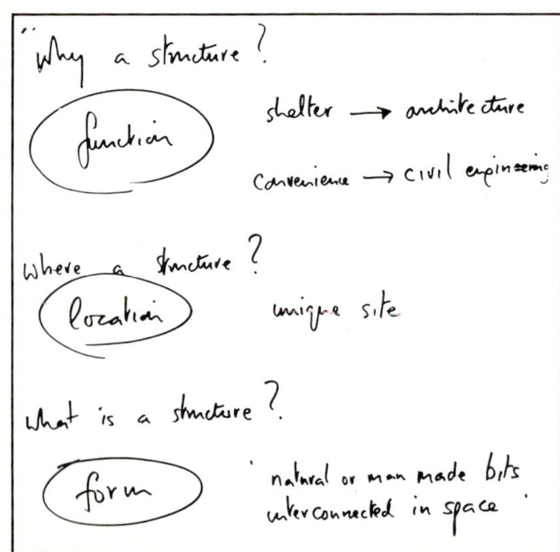

Why a structure?

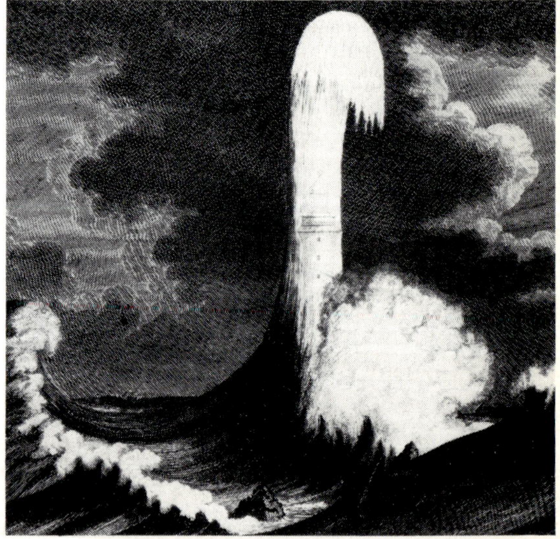
Unique site: Eddystone Lighthouse, 1759.

**What is a structure?** Structure may be defined as *natural or man-made bits interconnected in space*. Its essence and art lies in the understanding of how these bits work individually and together and in joining them in such a way as to form a stiff and stable whole. There is no science or law to say how the bits should be distributed in space. That is a decision made entirely by man, the engineer.

The creation of a structure from its conception to its physical presence has greatly changed with the increase in scientific knowledge and development of materials and technology. There is, however, a sequence of decisions to be made for a structure of any period.

To conceive a structure some knowledge of or instinct for stability is essential and this may come from an awareness of three-dimensional forms, from trees to primitive tents to skyscrapers. Stability is three-dimensional.

Loads are transmitted to the ground via various structural 'routes', each having its own stiffness. The stiffer the route the more the load is attracted and the distribution of stiffness in space is the art of the engineer.

The bits which make up a structure are subjected to tension, compression, bending and torsion and their action under load has always preoccupied engineers in their search for truth. Galileo, Hooke, Musschenbroeck, Coulomb and Girard are a few of the early men who studied the strength of materials.

Timber is the only natural tension material (though stone has some tensile strength) and its shaping and jointing became highly sophisticated long before the discovery of man-made tension materials



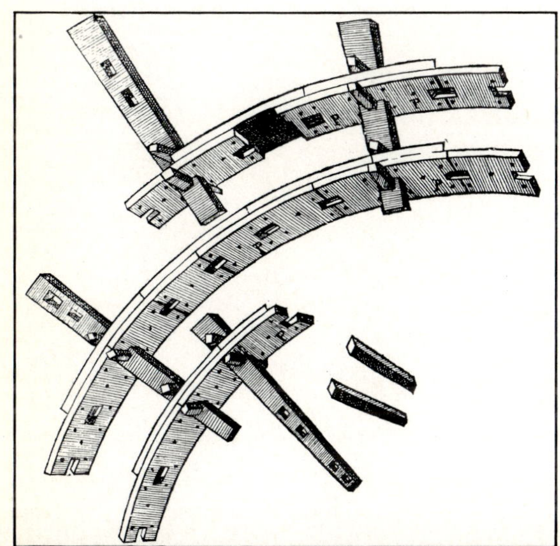

Bits in space: Philibert de L'Orme, 1561.

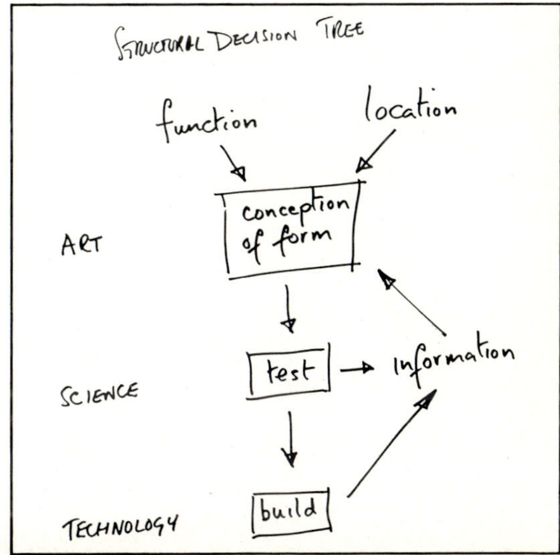


Structural Decision Tree.

4 such as cast and wrought iron, steel, aluminium and plastics. In areas where timber is not plentiful compression structures in stone and brick are traditional.

Each new material has its own manufacturing and jointing techniques which dictate shape and size of individual members. Timber is cut, cast iron poured into moulds, wrought iron and steel rolled, and aluminium extruded. Composite materials, such as concrete reinforced with steel or glass, or pre-stressed with high-tensile steel wires, have evolved as new materials capable of taking tension. (The first use of a new material or connection method often produces the most inventive structures before the fog of irrelevant data blurs its image).

Once proposed, a design must be checked for society by scientific methods which range from compliance with traditional rules of thumb, testing of component parts and connections, load tests on the whole structure, and mathematical analysis. For highly complicated structures model tests have now been overtaken by computer analysis.

Construction calls for skills of a different kind. Not only have the bits to be physically assembled in space and then connected but the works have to be kept stable at all times. The ease of construction reflects the engineer's awareness of practical problems of technology.

The engineer therefore builds up a store of information from his own experience, from scientific and technical books, reports and journals and from discussion with colleagues and friends in the profession.

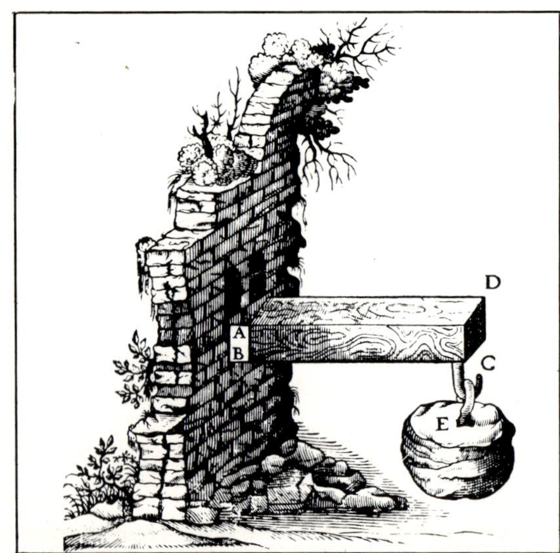

Testing: Galileo, 1638.

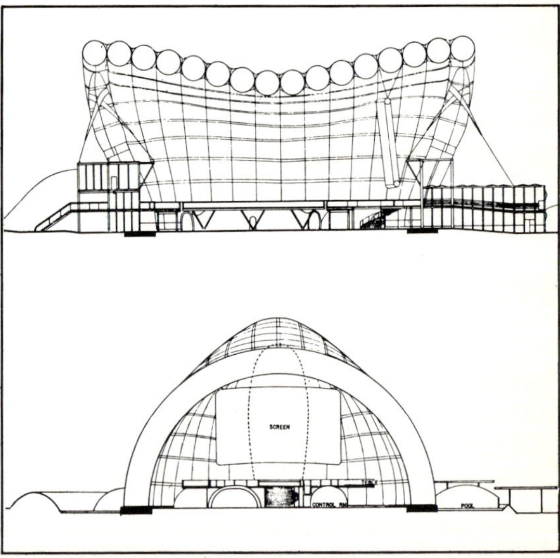



Concept of Form: Osaka Expo '70.

The basic framework or structural decision tree remains constant, whether applied to Perronet or to Samuely, while the volume of information has increased. It can thus be used to evaluate the achievement of any engineer at any time in history and is the basis of this exhibition.

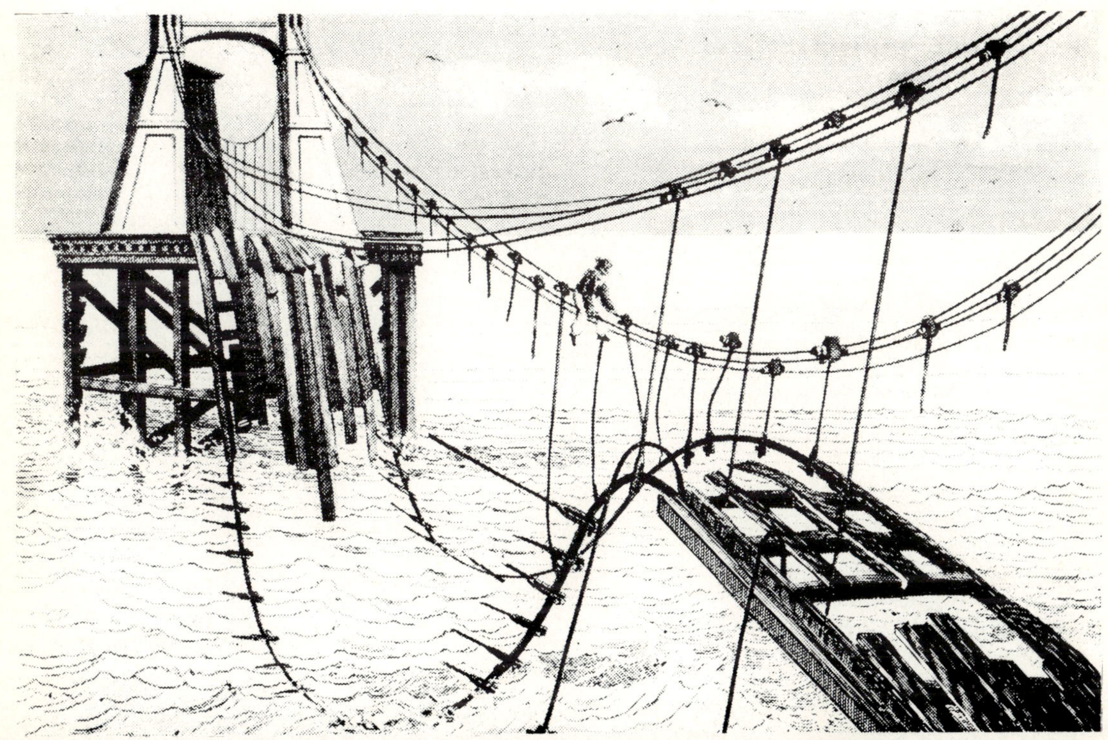

Information: Brighton Chain Pier after Storm Damage of 1833.

# JEAN-RODOLPHE PERRONET
## 1708-1794
### The Bridge at Neuilly sur Seine.

6
In 1772 the centering of Perronet's five-arch masonry bridge over the Seine at Neuilly near Paris, was lowered in the presence of the King amidst great celebration and publicity. In its construction Perronet, for the first time in bridge-building history, had effectively applied theory to practice, producing an entirely new architectural form.

Perronet was born in 1708 and at 17 became apprenticed to an architect, working successfully on both buildings and civil engineering projects until he was invited to join the Corps des Ponts et Chaussées by Trudaine, its founder. In 1747, when the Ecole des Ponts et Chaussées was started, he became its first director, passionately believing that theory and practice should be taught together. At the same time he continued his engineering duties, also assisting the ageing Hupeau, premier engineer of the time, on several projects including the Mantes bridge in 1757.

Traditionally bridges were built arch by arch, each of which had to be stable during construction, and design followed geometrical rules for the proportions of the various parts. These had originated from architects such as Alberti, Palladio and Blondel. From 1675, when Hooke published his theory of the inverted catenary, scientists had studied the behaviour of arches. The most widely accepted theory was that of La Hire and it was published, together with rules of thumb, in the earliest engineering handbooks; Gautier's 'Traité des Ponts' of 1716 and Bélidor's 'La Science des Ingénieurs' of 1729. These remained the standard reference books throughout the 18th century. Couplet's later theoretical work on the formation of hinges was confirmed by Danizy's experiments on plaster arch models and Poleni, when investigating the cracking of the dome of St. Peters in 1743, correctly applied arch theories to explain that it was fundamentally stable.

Perronet, during construction of the Mantes bridge, observed that in building the first arch, the supporting pier moved sideways with some cracking,

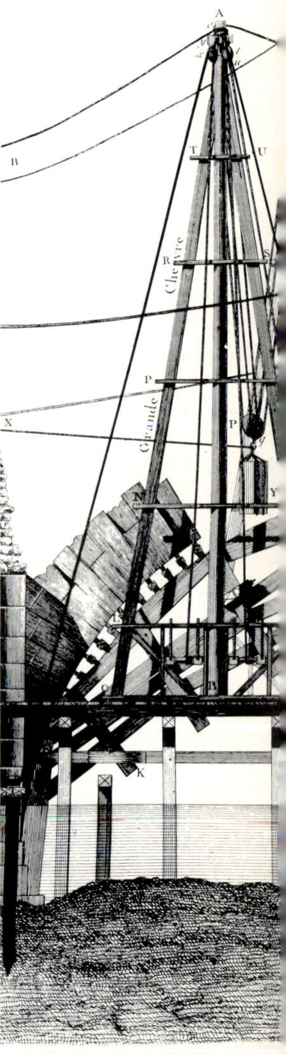

Centering for Mantes Bridge, 1757.

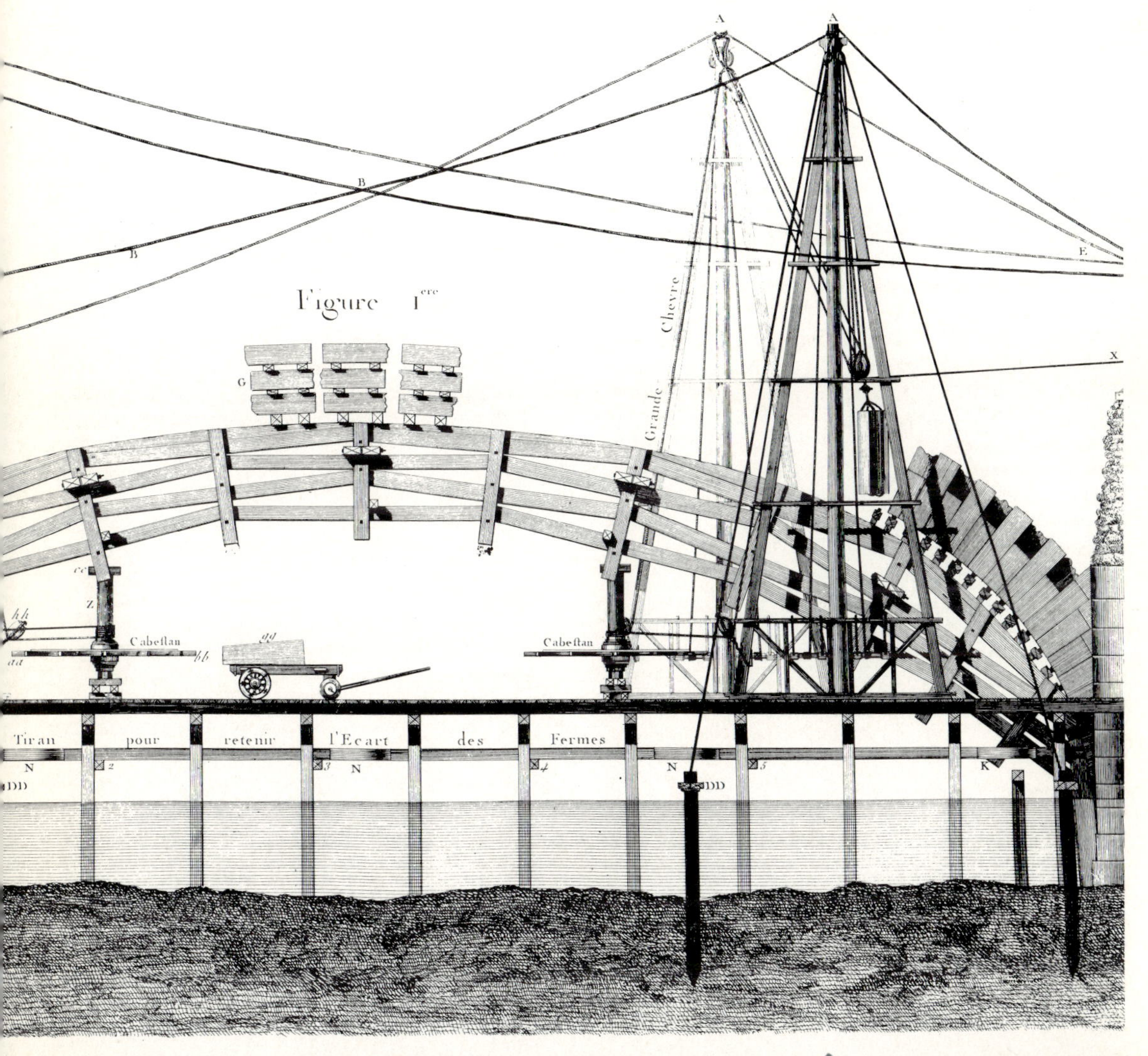
A
A
B
B
E
X
Figure 1ᵉʳᵉ
Grande Chevre
G
Cabestan
Cabestan
gg
hh
cc
z
aa
bb
Tiran pour retenir l'Ecart des Fermes
N 2
N 3
N 4
N
DD
DD
K
5

8 but when the second arch was built it righted itself and the cracks closed. He realised then that the piers only needed to be wide becaue they were required to carry the horizontal thrust during construction. In the completed multi-arch bridge the horizontal thrusts of adjacent arches balance each other and piers only carry the vertical load. Therefore, if the arches were built together and their centering struck at the same moment, the piers could be designed from the outset to carry vertical load only. Smaller piers and foundations would offer less resistance to the flow of water and provide considerable saving in costs of construction. Furthermore, Perronet also realised that the arches could be very much shallower if the abutments were adequate to withstand the resulting high thrust.

Before using this method of construction, he tested his theories of arch behaviour at Nogent, 1766-69, (thereby reinforcing Danizy's earlier model tests of 1732) and with the confirmation of his theories he had the confidence to design and build Neuilly with shallow arches and slender piers.

With the principle established a new era in bridge building began. His next bridge at Pont-Sainte-Maxence (1774-85) is the most radical of all, which, compared to traditional structures, had an increase in span to rise ratio from 3 to 11; today's bridges are of similar form.

Perronet, the father of modern bridge building, published many reports throughout his career, culminating with 'Description des Projets' of 1782/83, which were widely studied. His friend and fellow engineer, Lesage, wrote after his death:

*Comme ingénieur, Perronet s'est constamment occupé, pendant sa longue carrière, des connoissances théoriques qu'il est possible d'acquérir; de l'art de concevoir les projets en grand, et de les bien rédiger; enfin, de celui des grandes constructions subordonnées aux règles de la statique et aux principes du bon goût, et d'une sage économie.*

## PERRONET · EXHIBITS

A1    Jean-Rodolphe Perronet. 'Description des Projets et de la construction des ponts de Neuilly, de Mantes, d'Orléans & autres...', 1782/83. Engraved portrait.

A2    Perronet, 'Descriptions des projets...'. Plate XI, ceremonial decentering, on 15 September 1772, of the Neuilly bridge.

A3    H. Gautier, 'Traité des ponts', 1716.

A4    Bernard Forêt de Bélidor, 'La science des ingénieurs dans la conduite des travaux de fortification et d'architecture civile', 1729.

A5    M. Frezier, 'La théorie et la pratique de la coupe des pierres et des bois... ou traité de stéréotomie', 1739. Plate III in Vol 3 showing Danizy's model tests on arches carried out in 1732.

A6    G. Poleni, 'Memorie istoriche della gran cupola del tempio vaticano', 1748. Plate XV, internal view showing cracking.

A7    Poleni, 'Memorie istoriche...', 1748. Plate E, showing determination of line of thrust by experiment.

A8    Poleni, 'Memorie istoriche...', 1748. Plate H, showing possible mode of failure.

A9    Perronet, 'Description des projets...', 1782/83. Plate 22, Mantes bridge under construction.

A10    Perronet, 'Description des projets...'. Plate 68 (from the Supplement of 1789) showing the measurement of the arch movement on the bridge at Nogent.

A11    Perronet, 'Description des projets...', 1782/83. Plate VII, profile and progress of works at Neuilly, end of 1770/71.

A12    Perronet, 'Description des projets...', 1782/83. Plate XVII, showing Perronet's waterwheels driven by the river current.

A13    Perronet, 'Description des projets...', 1782/83. Plate XXVII, showing centering.

A14    Perronet, 'Mémoire sur la recherche des moyens que l'on pouroit employer pour construire de grandes arches de pierre de deux cents, trois cents, quatre cents, et jusqu'à cinq cents pieds d'ouverture qui seroient destinées à franchir de profondes vallées bordées de rochers escarpés', 1793.

A15    Perronet, 'Observations du Sieur Perronet sur le mémoire remis le 17 Avril 1764 en faveur des crèches des ponts, et contre les grandes arches', 24 April 1764'.
Manuscript in Perronet's hand (Frank Newby).

A16    Perronet, 'Mémoire sur les différentes méthodes qui ont été employées pour fonder les ouvrages de mâconnerie dans l'eau', 13 Nov. 1765. Contemporary manuscript copy (Frank Newby).

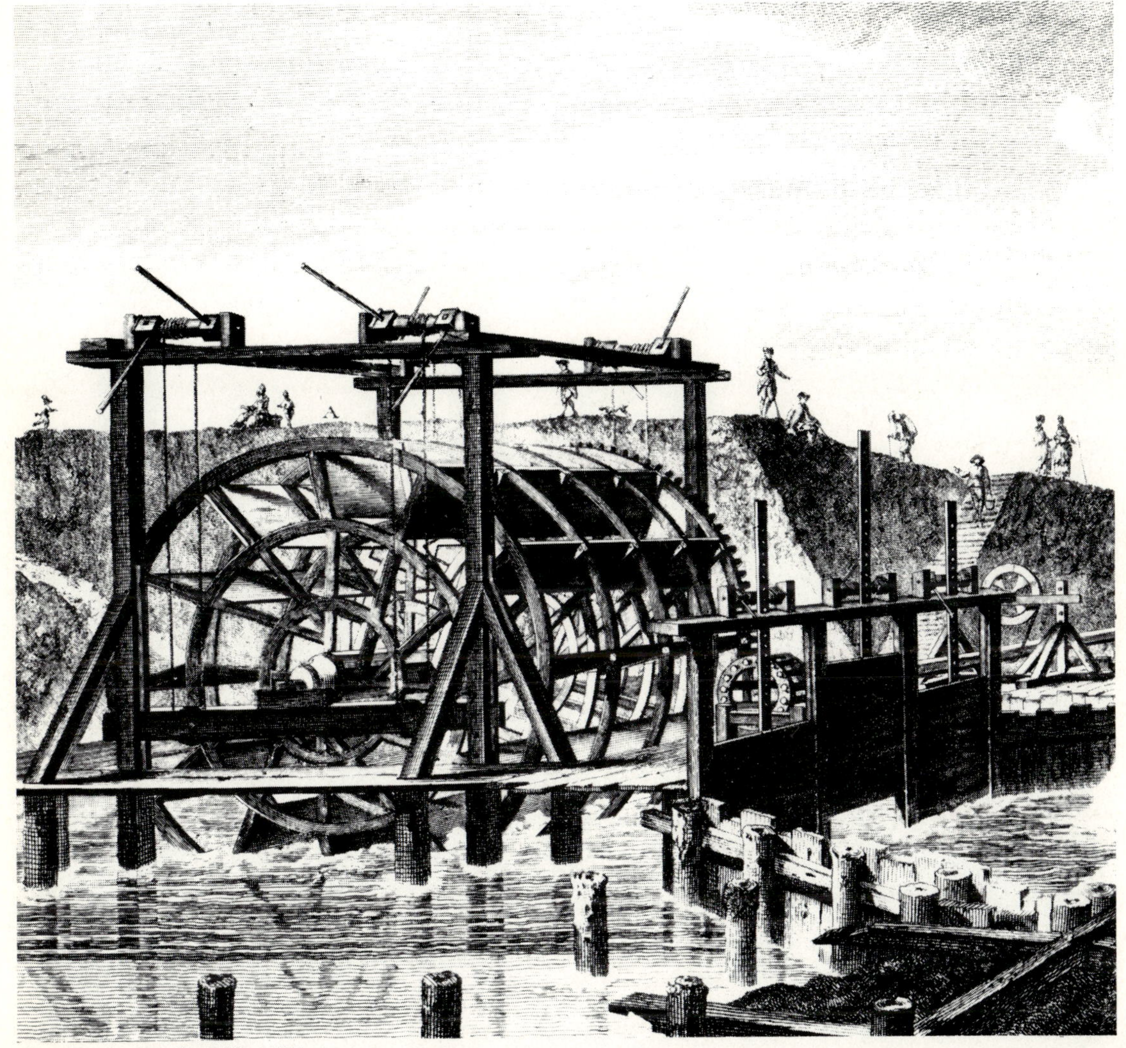

Waterwheels at Neuilly Bridge, 1768.

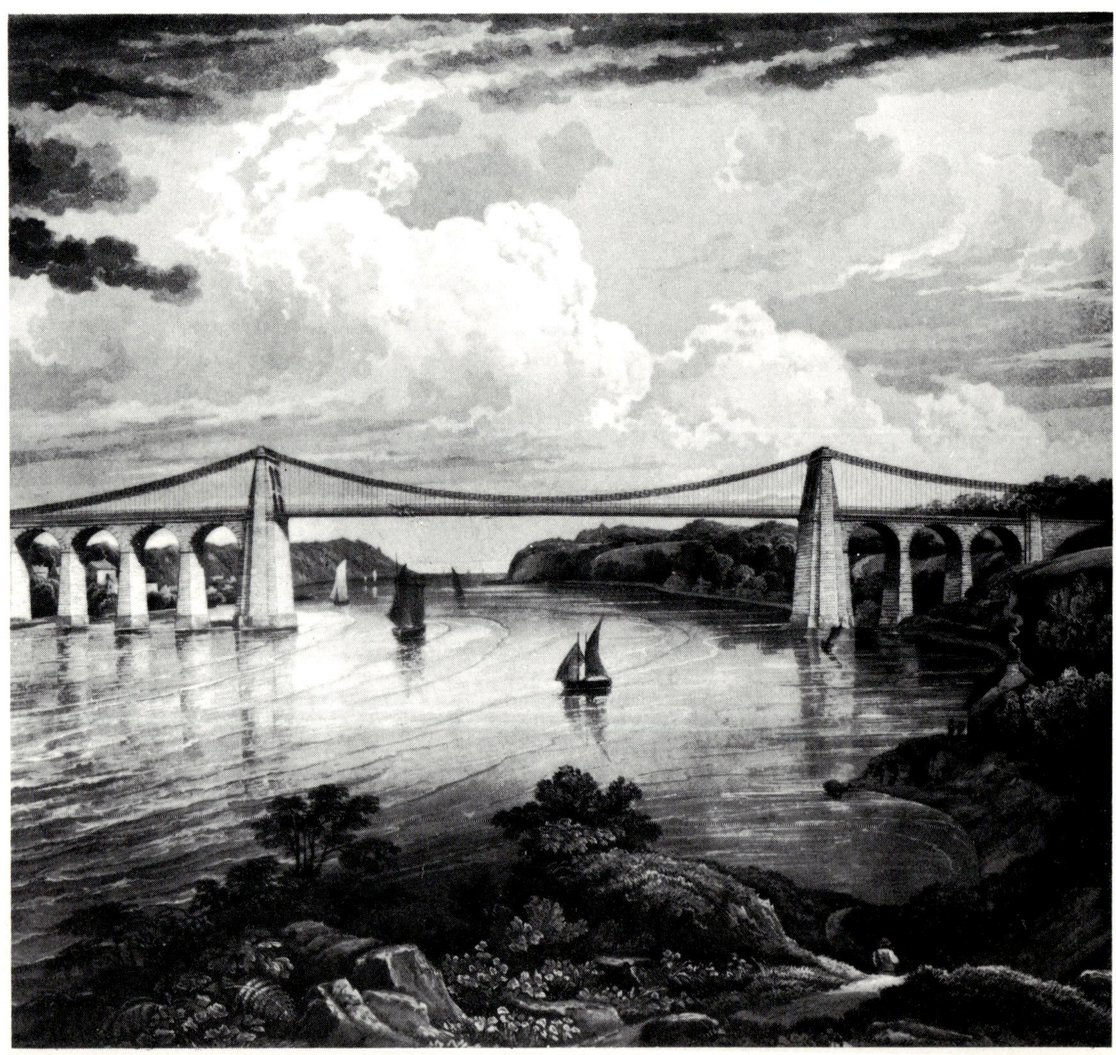

Menai Suspension Bridge, 1826.

# THOMAS TELFORD
**1757-1834**
**London Bridge Design 1800-1801**

London's transformation into Britain's major port began at the very end of the 18th century with the construction of London and West India docks. The Select Committee upon the Improvement of the Port of London was also seriously considering the demolition and replacement of old London Bridge as part of its radical plans to relieve the congested river. They held an open competition, finally deciding that they wanted a bridge of iron, then a new and exciting material, with headroom of 65ft above high water to acommodate ships of up to 200 tons burden. Entries were published in July 1800 and included designs for multi-span iron bridges by Telford and his assistant, James Douglass, and by Thomas Wilson. Wilson was the only engineer with any experience in large-scale cast-iron construction since he had already built the Sunderland Bridge with its very long span; his was originally the preferred design for London Bridge. However, at the end of 1800 and too late for inclusion in the Committee's report on the competition entries, Telford submitted his bold and revolutionary scheme for a single cast-iron arch with the enormous span of 600ft.

In conceiving a structure of this magnitude in a relatively untried material, Telford was influenced by several factors. In 1787 he was appointed County Surveyor of Shropshire where, less than ten years before, Abraham Darby III had cast the world's first iron bridge in his Coalbrookdale Foundry and erected it over the River Severn nearby. Telford was on terms of friendship with many of the Shropshire ironmasters, including Darby, and his own first iron bridge at Buildwas was cast at Coalbrookdale in 1796. He also knew Charles Bage who in 1796 built one of the earliest mills to use cast-iron columns and who made a valuable contribution to the knowledge of the strength of the material. But the man who was to have most effect on him was William Reynolds of the Ketley Ironworks whose family had been in partnership with the Darbys. In 1795 Telford and Reynolds collaborated over the iron aqueduct at Longdon on Tern (the first iron structure with which Telfore is associated). The great aqueduct at Pontycysyllte (1802-5) with its iron trough 1007ft long evolved from Longdon, for the experience Telford gained by working with this great iron founder was directly responsible for his subsequent mastery of cast iron as an effective structural material.

But of particular significance for the London Bridge design was the erection, in 1796, of Wilson's famous iron arch bridge over the River Wear at Sunderland with its unprecedented single span of 236ft. Unlike Ironbridge and Buildwas whose arch ribs were cast in two or three very large pieces, the arch ribs at Sunderland were built up of smaller interconnected cast-iron elements, like voussoirs in a masonry bridge. This idea first appeared in France in various unexecuted schemes, notably those of Racle and Monpetit, and these were known in England. The very fact that a bridge on the scale of Sunderland could be successfully built greatly influenced Telford.

When Telford submitted his proposal it therefore aroused considerable interest and the Select Committee were also attracted by the advantages to navigation of a single arch. They were anxious however, to ascertain the feasibility of such a project and Telford prepared a series of questions with a further drawing showing latticed spandrel bracing, to be circulated amongst the best engineers, ironfounders and scientists of the day. These were published, together with the replies, in a report of 3 June 1801. The ironmasters considered that the cast-iron bits could be made, though some testing of individual elements should be carried out, and Wilkinson even advised a large-scale model test of the whole arch. However, the report revealed that there were too many unknown factors involved, not least in the foundation stability of the abutments which would

12 have rested on gravel a few feet above the London clay, rather than on rock. These uncertainties, coupled with the considerable planning problems of building the high-level approaches, meant that the Committee were unable to make any definite recommendations and the bridge remained unbuilt.

But the project was not to be in vain. The report of 1801 exactly reflects the state of contemporary knowledge of both the material and of arched bridges and Telford himself wrote of it that it would be 'the means of throwing much new light on this important subject and will probably change the principles and practice of this species of architecture'. In 1811 Telford produced his definitive cast-iron arch bridge design which he used first at Bonar and immediately afterwards at Craigellachie, culminating in 1826 with the finest of them all at Tewkesbury. The latticed spandrel bracing which first appeared in the 1801 report was refined by Telford to become the most distinctive feature of these bridges. This form is still in use today.

## TELFORD · EXHIBITS

B1    Portrait of Thomas Telford after the painting by Raeburn. Mezzotint engraving (The Institution of Civil Engineers).

B2    Wilson Lowry after Thomas Malton, 'Perspective view of the design for a cast iron bridge, consisting of a single arch 600 feet in the span and calculated to supply the place of the present London Bridge'
Hand coloured aquatint, 1802 (The Institution of Civil Engineers).

B3    Thomas Telford and James Douglas, 'Messrs Telford and Douglass's design of a cast iron bridge of a single arch proposed to be erected across the River Thames'.
Engraving, 28th July 1800.

B4    'Mr. T. Wilson's design for a cast iron bridge over the Thames, instead of the present London Bridge'. Plate 37 of 'Reports from Committees of the House of Commons, Vol XIV Miscellaneous Reports'; also Port of London, with engraving, 1793-1802', 1803. 'Originally Plate VIII of 'Third report from the select committee upon the improvement of the Port of London', 28th July 1800, reprinted in toto though reduced in size, in 1803.

B5    'Messrs. Telford and Douglass's explanatory drawings'. Plate S4 of 'Reports from committees... vol. XIV. Originally plate.

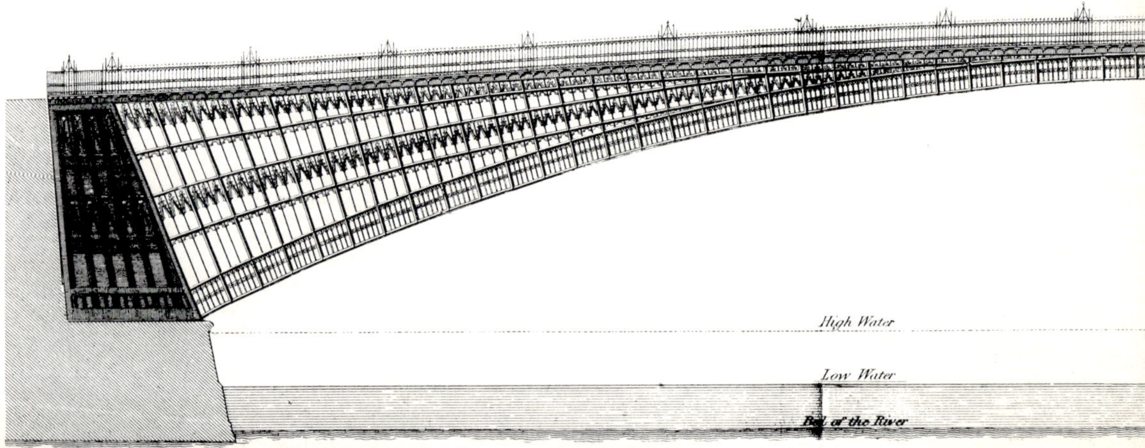


Telford's London Bridge Design, 1800

XXV/I of 'Fourth Report from the select committee....', 3rd June 1801, reprinted in toto, though reduced in size, in 1803.

B6 Joseph Plymley, 'General view of the agriculture of Shropshire', 1804.
Plate 3, 'Perspective view of a part of the iron aqueduct which conveys the Shrewsbury Canal over the River Tern at Longden in the County of Salop' (Ironbridge Gorge Museum Trust, Telford Collection).

B7 Plymley, 'General view of the agriculture of Shropshire', 1804. Plate 4 'Plan, elevation and section of the iron bridge built over the River Severn at Buildwas in the County of Salop in the years 1795 and 1796 (Ironbridge Gorge Museum Trust, Telford Collection).

B8 W. Hay, 'Bridge proposed to be erected over the River Spey at Craig-Elachie between the Counties of Banff and Elgin. Watercolour c.1812 (B. Weinreb Architectural Books Ltd).

B9 Thomas Telford, 'Life of Thomas Telford, Civil Engineer, written by himself... with a folio atlas of copper plates', 1838. Plate 81 from atlas. 'Cast iron bridge over the River Severn at Tewkesbury in the County of Gloucester'.

B10 'Transactions of the Institution of Civil Engineers', 1838, Vol 2 Plate III. Details of Tewkesbury Bridge.

B11 Vincent de Monpetit, 'Prospectus d'un pont de fer d'une seule arche', 1783.
John Rennie's own manuscript (Frank Newby). Copy of the plate from the printed work.

B12 Cast-iron Coalbrookdale fire-place with the Iron Bridge (Ironbridge Gorge Museum Trust).

B13 A set of Telford's drawing instruments (The Institution of Civil Engineers).

B14 Telford silver medal awarded by the Institution of Civil Engineers as a prize with the head of Telford on the obverse face and the Menai Bridge on the reverse.

B15 J. Raffield after Robert Clarke, 'West view of the cast iron bridge...over the River Wear at Sunderland in the County of Durham'.
Aquatint c. 1796 (Ironbridge Gorge Museum Trust, Elton Collection).

B16 Marc Brunel. Unpublished manuscript report on Sunderland Bridge, 1 June 1818 (B. Weinreb Architectural Books Ltd).

B17 A pair of Sunderland lustre jugs, one showing Thomas Wilson's bridge and one showing it after it was completely rebuilt in 1858-9 by Robert Stephenson.



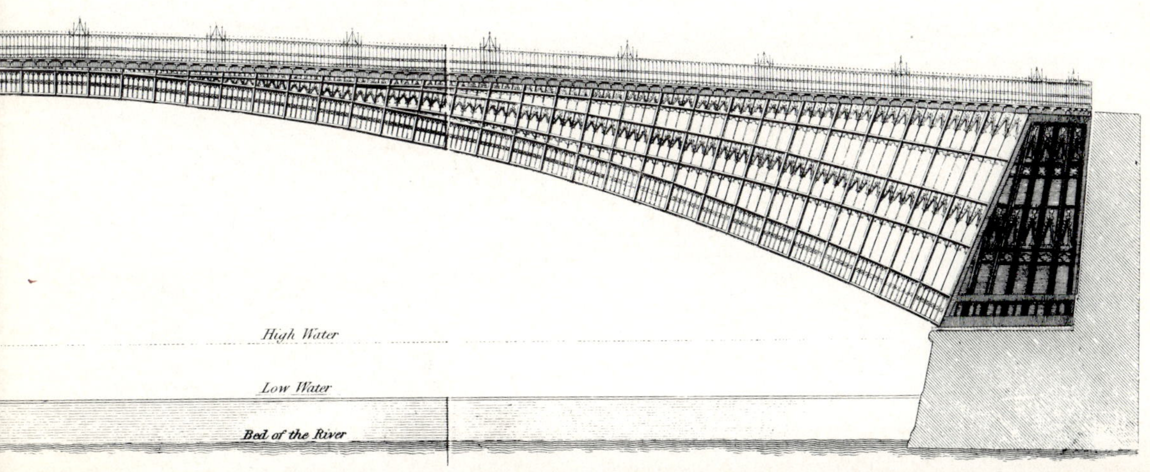

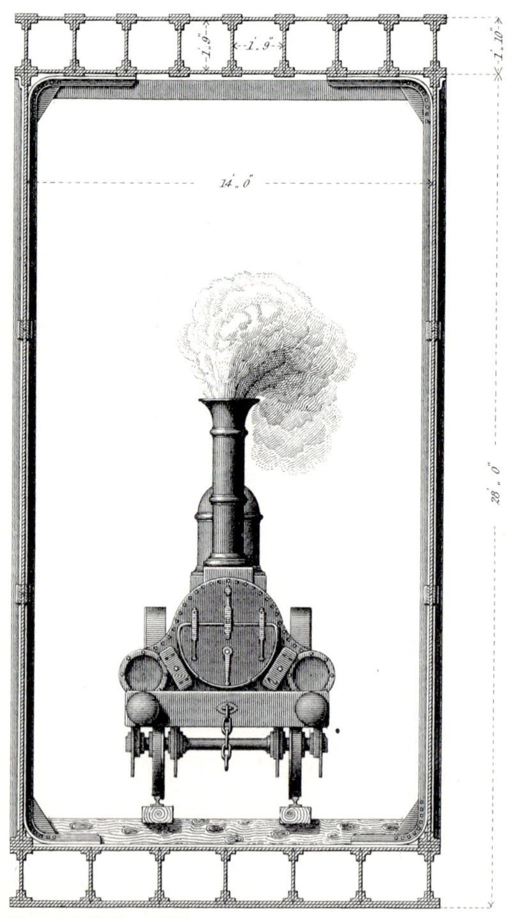

Britannia Bridge, 1850.

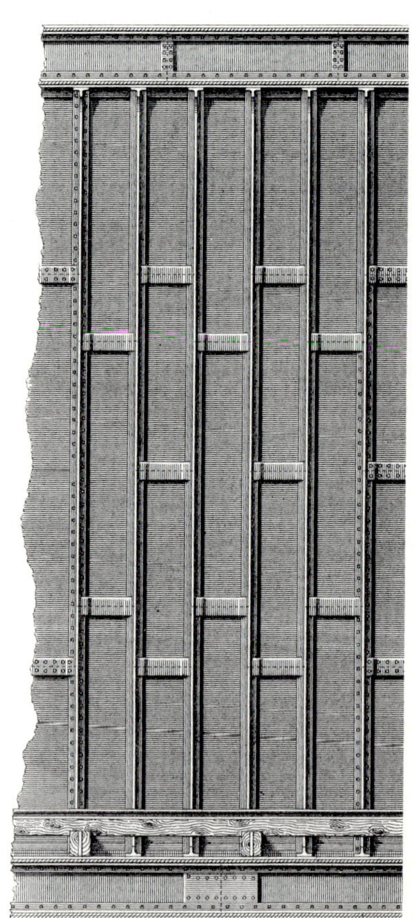

# ROBERT STEPHENSON
**1803-1859**
**Britannia Tubular Bridge**



In 1844 the Act of Parliament was passed authorizing the building of the Chester and Holyhead railway to complete the rail link from London to Dublin via Holyhead on Anglesey. Robert Stephenson was appointed engineer and almost at once turned his attention to the vital question of how to bridge the Menai Straits.

Telford before him had faced the same problem when building the Holyhead Road, for in both cases the Admiralty forbade the closure of the Straits at any time during construction. Telford had submitted a design for a cast-iron arch of 500ft span to be erected by a newly devised method to avoid obstructing the waterway. However, the headroom provided was not acceptable to the Admiralty. He finally solved the problem with his suspension bridge of 580ft span (130ft longer than any hitherto) completed in 1826, crossing the narrowest portion of the Straits.

When Stephenson came to consider the same problem, he sited the bridge at a point where a rock in the middle of the Straits offered foundation for a central pier so that the maximum span would be limited to 460ft He originally proposed a cast-iron arch bridge but the Admiralty as before refused it, insisting on 100ft headroom. Stephenson also knew that a suspension bridge lacked adequate stiffness to support the very heavy moving load of a locomotive.

Given the unprecedented lengths of the spans to be bridged Stephenson from the outset considered the idea of an immense hollow tube through which the train would run. The tube was to be built up of wrought iron plates, rivetted together in a manner similar to that of ship construction, and Fairbairn, a master iron shipbuilder, expressed confidence in the idea.

Since wrought iron had never before been used to make up a long-span structure, there was no available information on which to draw. Fairbairn, together with Hodgkinson, whom he had invited to join the team, began a series of model tests to determine the best shape of tube and to investigate its structural behaviour. They established that the tubes should be rectangular rather than circular or elliptical and that cellular sections be incorporated on the top and bottom flanges to prevent these from buckling or rupturing and a structural design method was evolved which could be applied to all forms of beams. This synthesis of theory and practice marks the advent of structural engineering as we know it today.

In order to keep the Straits open, the central tubes, each 460ft long, were fabricated on the shore. They were floated on pontoons at high tide into position between the piers and raised into place, 100ft above the water, by means of hydraulic jacks. The bridge was opened in 1850.

The Britannia tubular bridge was the most famous bridge of its age and detailed accounts of it from conception to completion were written by Edwin Clark, Stephenson's resident engineer, and by Fairbairn. The all-important theoretical work was also included in the remarkable report on the Application of Iron to Railway Structures produced as a result of Stephenson's Dee bridge failure in 1847. This document contains the opinions and experience of the foremost engineers, scientists and iron manufacturers of the time, and is a unique record of the art of iron bridge building at the very moment when cast iron was giving way before the overwhelming superiority of wrought iron to provide for the demands of the railways.

## STEPHENSON · EXHIBITS

C1  Samuel Bellin after John Lucas. Full length portrait of Robert Stephenson. Mezzotint engraving, 1753.

C2  View of the Britannia Bridge and Telford's suspension bridge. 19th century Welsh straw picture.

C3  S. Russell. View showing one of the tubes under construction. Coloured lithograph, 1848 (Ironbridge Gorge Museum Trust, Elton Collection).

C4  G. Hawkins. View showing the second tube being floated out. Coloured lithograph, 1849 (Ironbridge Gorge Museum Trust, Elton Collection).

C5  Autograph letters from Robert Stephenson to Captain Moorsom, 10 March 1848, 5 May 1849 (The Science Museum).

C6  Edwin Clark, 'The Britannia and Conway tubular bridges', 1850. Plate 1 from atlas. General plan of works and operation of floating the first tube.

C7  Clark, 'The Britannia and Conway tubular bridges', 1850. Plate 13 from atlas. Isometrical projection of a portion of one of the tubes.

C8  Clark, 'The Britannia and Conway tubular bridges', 1850.

C9  William Fairbairn, 'An account of the construction of the Britannia and Conway tubular bridges', 1849. Plate VI. Longitudinal and transverse section of part of the Britannia tubes.

C10  Fairbairn, 'An account...'. Plate VII. Perspective view of a portion of the Britannia tubes.

C11  A group of autograph letters from William Fairbairn to Robert Stephenson, 1845-1847 (The Institution of Civil Engineers).

C12  John Grantham, 'On iron shipbuilding with practical examples and details', 1858. Plate 12, section of ship's hull.

C13  Thomas Jackson, 'The tourist's guide to Britannia bridge', 1854 (12th edition).

C14  L.Yvert, 'Notice sur les ponts avec poutres tubulaires en tôle', 1851.

C15  'Copy of report to the commissioners of railways, by Mr. Walker and Captain Simmons, R.E., on the fatal accident on 24th day of May 1847, by the falling of the bridge over the River Dee, on the Chester and Holyhead railway...', 1847.

C16  'Report of the commissioners appointed to inquire into the application of iron to railway structures', 1849.

C17  'Cast iron bridge 121′ 6″ between the bearings erected by H & M.D. Grissell Regents Canal Iron Works London'. Watercolour c. 1850 (B. Weinreb Architectural Books Ltd). The original for plate 42 in the atlas of Clark's 'The Britannia and Conway tubular bridges'.

C18  Commemorative medal struck for the opening in 1850 of the Britannia tubular bridge with head of Robert Stephenson on obverse and view of the bridge on reverse.

C19  Corner piece of one of the top cells taken from the bridge during demolition.

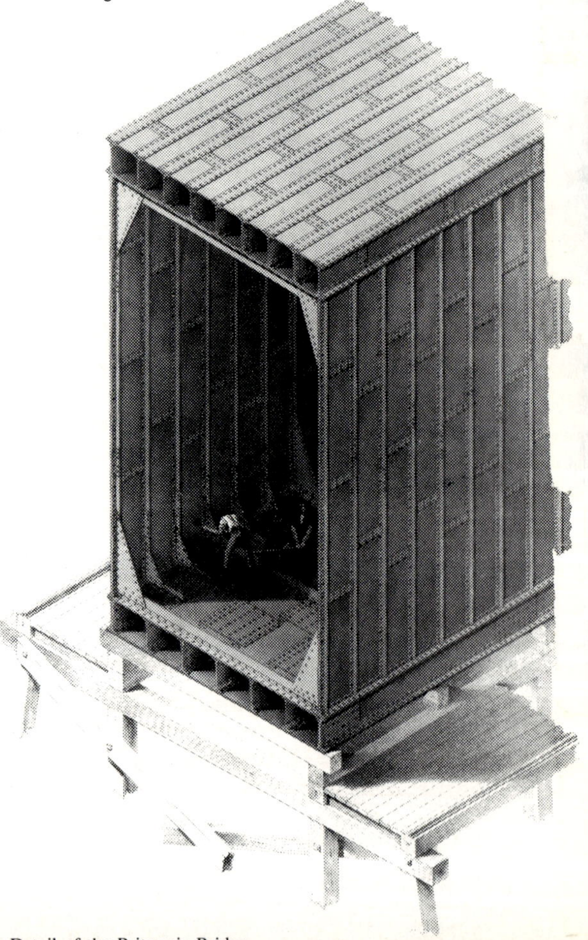

Detail of the Britannia Bridge

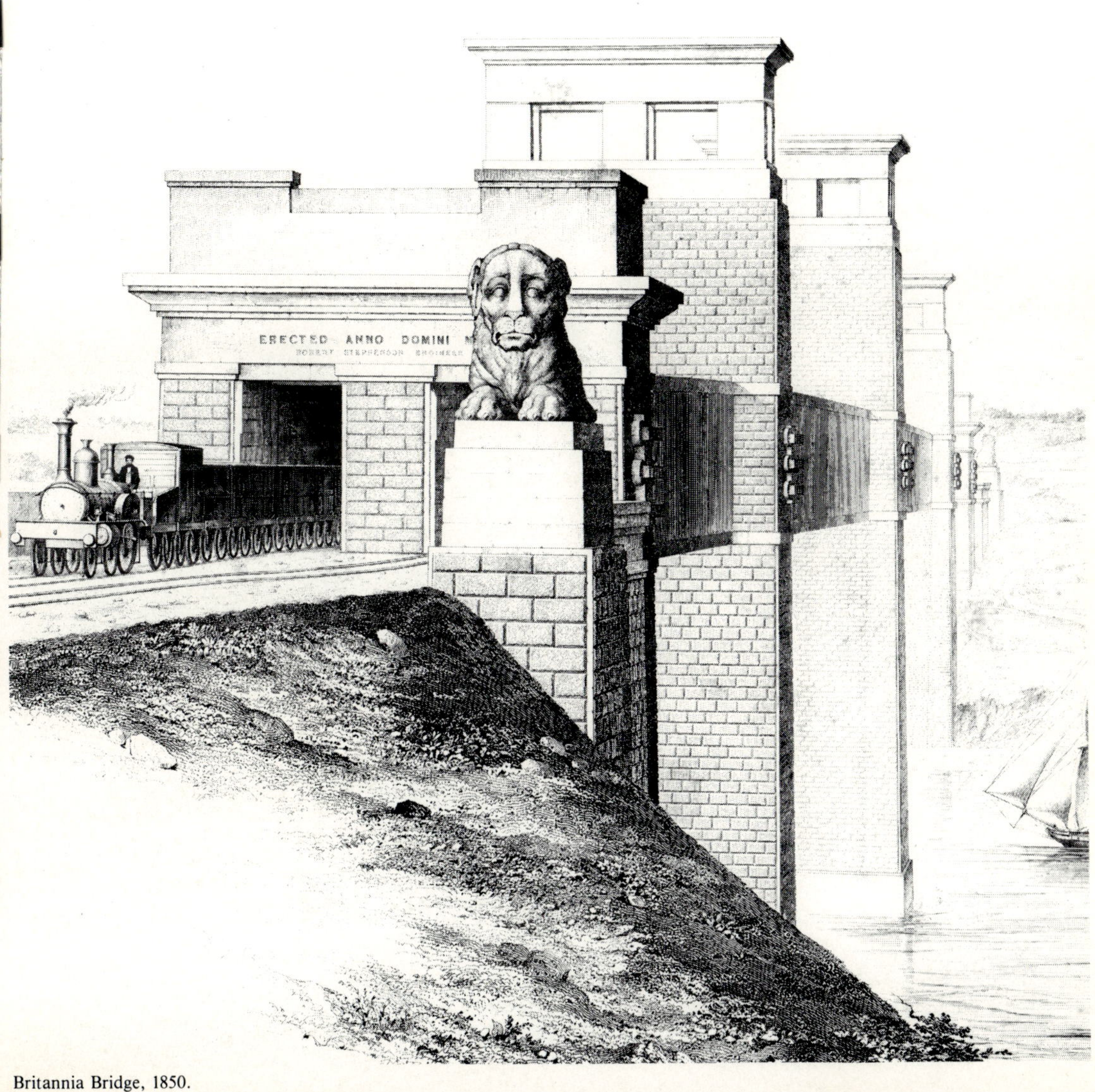


Britannia Bridge, 1850.

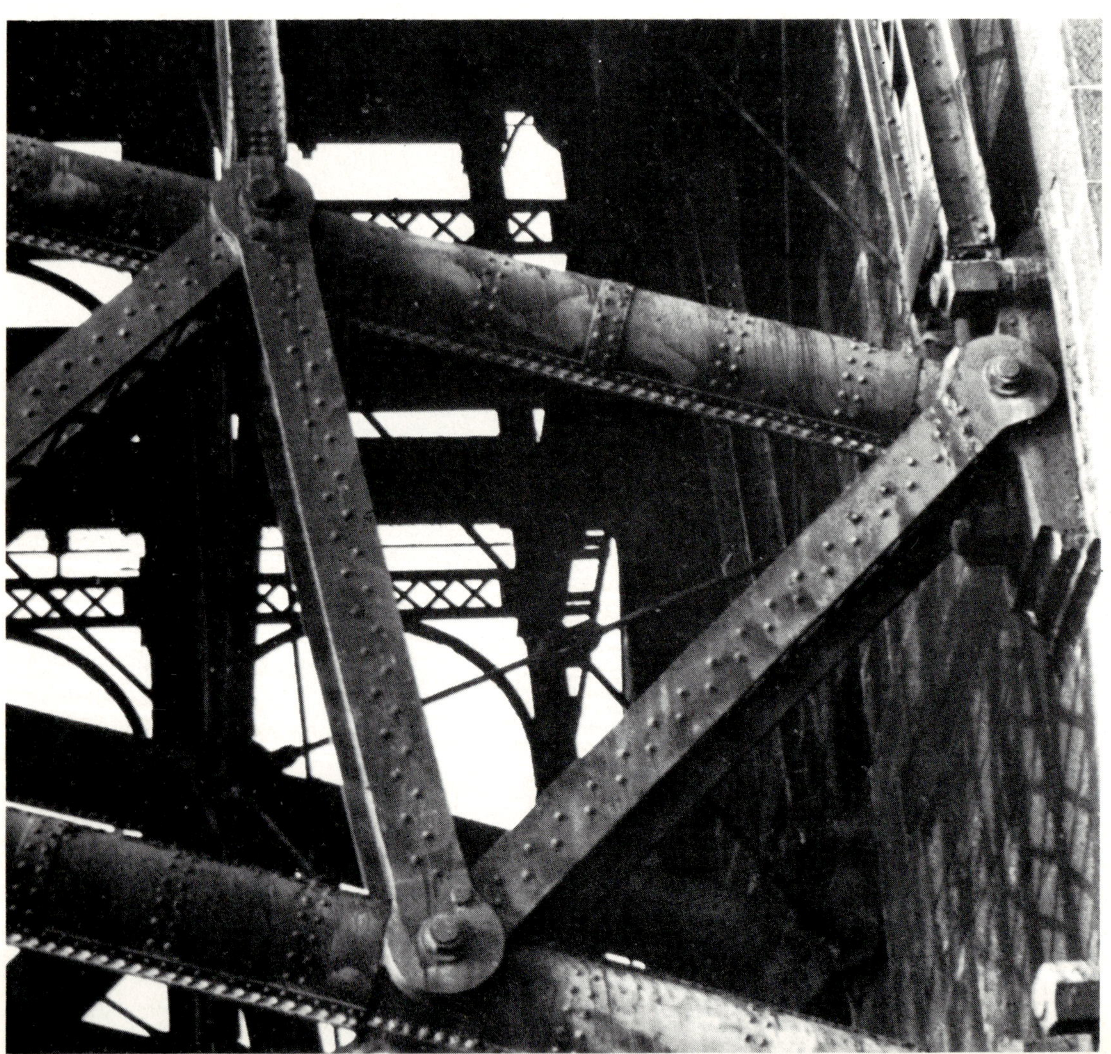

Detail of St. Louis Bridge.

# JAMES EADS
**1820-1887**
**St. Louis Bridge**

In 1856 the railroad arrived at the east bank of the Mississippi river at St. Louis, 1500ft wide at this point. The problem of bridging such a crossing had been considered earlier. Charles Ellet in 1839 had submitted a proposal for a suspension bridge to carry road traffic and in 1855 Dent and John Roebling had each proposed a railway suspension bridge. In 1856 the city engineer put forward plans for a tubular bridge similar to Stephenson's Britannia bridge with foundations going all the way down to rock. However the matter rested until 1865 when the St. Louis and Illinois Bridge Company was founded.

It was decided to build a combined road and rail bridge with either two spans of 350ft or one of 500ft so as not to obstruct the river traffic. Furthermore, because of the large number of suspension bridge failures this type of structure was ruled out.

Senator Brown, who was responsible for getting the bill approving their plans through Congress, opened it with the following statement, clearly demonstrating the importance which was attached to the project:

'I want the structure when built to be one worthy of the great States it is to connect, of such ample capacity as will permit the freight of all the railway lines that may hereafter center at the great distributing point of the continent. Let it be built, too, for the ages, of a material that shall defy time, and of a style that will be equally a triumph of art and a contribution to industrial development.'

After much local controversy James Eads presented his remarkable arched bridge design in 1867 which was eventually accepted and he became chief engineer of the Illinois and St. Louis Bridge Company. Eads was born in 1820 in Indiana, was self-taught in mechanics and started work on the river with a steam boat at the age of 19. Although he had never before built a bridge he had an unsurpassed knowledge of the river bed from his successful underwater salvage business. He had travelled widely in the United States and in Europe, seeing, for example, the newly-completed iron arch railway bridge at Coblenz. He would also appear to have read widely, quoting Telford and Brunel.

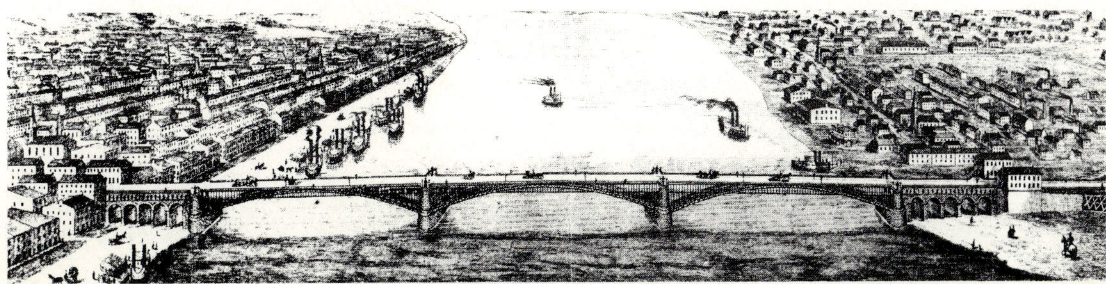

Overview of Eads's proposed bridge at St. Louis, design of 1867.

For the superstructure of his own bridge he proposed three lattice-steel arches with a central span of 500ft. He appreciated the potential of steel and with a fresh mind came up with a detailed design of great inventiveness. He saw the bridge not only in terms of public elegance but also in terms of technical innovation.

In the presentation of his scheme he made some telling remarks, 'If there were no engineering precedent for 500 foot spans can it be possible that our knowledge of the science of engineering is so limited as to teach us whether such plans are safe and practicable? Must we admit that because a thing has never been done, it can never be, when our knowledge and judgement assure us that it is entirely practical? This shallow reasoning would have defeated the laying of the Atlantic cable, the spanning of the Menai Straits…and left the terrors of the Eddystone without their warning light!'

His first report of 1868 included the following: 'Fully estimating the great responsibility assumed in undertaking to design and complete this important work (it is necessary to) secure the aid of the ablest talent…and (prove) by careful experiment, as far as possible, everything connected with it'. Eads on his appointment had taken into employ two German emigrés, Flad and Pfeifer (who had written a prize thesis on arched bridge design). They carried out calculations which were checked by Professor Chauvenet. Of the steel members Eads writes, 'To ensure a uniform quality and high grade steel at the lowest prices and at the same time avail myself of the advantages of the tubular form of construction, I propose to have the steel rolled for the arches in bars of 9 feet length and of such form that ten of them shall fit the circumference of a lap welded tube one eighth of an inch thick in the manner that staves of a barrel fill the hoops. This would virtually form a steel tube 9 inches in diameter and of 6 inches bore…'. Eads specified that only steel bits with the same stiffness be used to make up the tubes, thus eliminating any uneven distribution of load within the tube. As the design progressed some simplifications were made but the form and concept remained the same.

From the beginning Eads felt that the foundations of the bridge should go right down to rock as he was well aware of the very fast current at the bottom of the river and the consequent scouring effect of the sand carried with it (because of this he had considered the proposed pile foundations of the earlier schemes to be inadequate). Borings had shown that sand, varying in depth from 15 to 100 feet, overlaid the limestone bedrock and the cofferdams he intended to use would have to be constructed at a deeper level than any previously.

The shallow west abutment was started in 1867 and built within a traditional cofferdam. While this was underway, Eads was again in Europe and during his visit he discussed his plans for the very deep piers and eastern abutment with some of the leading French and British engineers. He was also able to see the pneumatic method of constructing underwater foundations. He realised that this would be a far superior way of sinking his piers to the great depths required than using the cofferdams he had originally intended. At St. Louis, therefore, he made the first large-scale use of compressed air caissons, contributing not only to the advance of engineering science but also to medical science, for the observations that he and his resident doctor made on caisson disease laid the basis for the present-day compressed air safety laws.

In 1873 erection of the superstructure finally began. There had been long delays owing to Eads's stringent requirements for the exceptionally high-strength steel members and the inability of Carnegie and his manufacturing company to supply them from carbon steel. However the new standards in quality

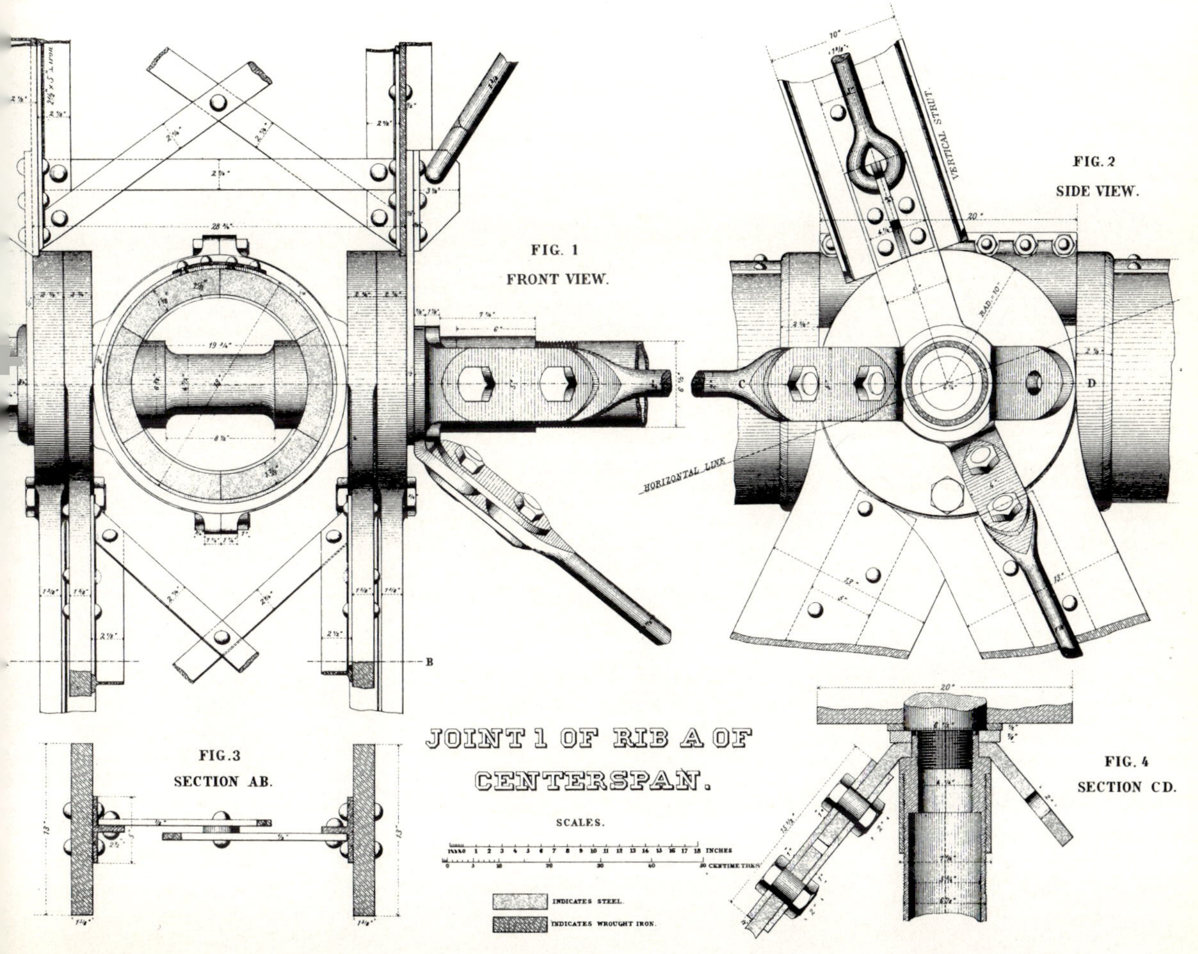


Joint Detail of St. Louis Bridge.

22 control and workmanship which evolved in this period greatly benefitted Carnegie and the emerging steel industry. At the time it was found that chrome steel, supplied by another company, had the required strength but Carnegie continued to supply the wrought iron.

The method of erection, devised by Henry Flad, was of interest for its similarity to an idea by Telford of 1818 for his proposed cast-iron arched bridge at Menai. It eliminated the need for staging in the river.

## EADS · EXHIBITS

D1    C.M. Woodward, 'A history of the St. Louis Bridge', 1881. Engraved portrait of Eads.

D2    Georg C. Mehrtens, 'A hundred years of German bridge building', 1900.
Rhine bridge on the Coblenz-Lahnstein railway line.

D3    Contemporary photograph of the Water-Pipe bridge, Washington aqueduct.

D4    Illinois and St. Louis Bridge Company, 'Report of the Chief Engineer', October 1870.

D5    Detail of Eads's original design of 1867 (from a plate in 'Engineering', October 1868).

D6    Woodward, 'A history of...', 1881. Frontispiece of the completed bridge.

D7    Woodward, 'A history of...', 1881. Plate XIII showing a pneumatic caisson.

D8    Woodward, 'A history of...', 1881. A group of plates showing connections of members, couplings and joints.

D9    Quinta Scott, 'The Eads Bridge. Photographic Essay', 1979. 2 photographs with details of joints and couplings.

D10    Woodward, 'A history of...', 1881. Plate XLIII, erection of west arch.

D11    Thomas Telford, 'Life of Thomas Telford, civil engineer, written by himself... with a folio atlas of plates', 1838.
Plate 77, design for the suspended centering for the proposed iron arch over the Menai Straits.

D12    Merchants Exchange of St. Louis, Cert. of Membership, 1882.

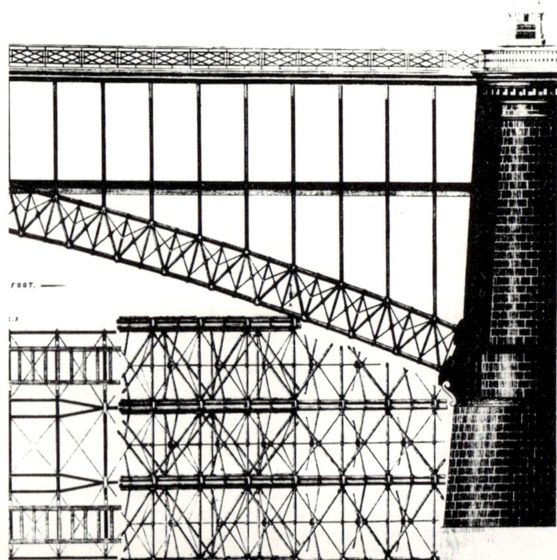


Detail of original, design of 1867.

# GUSTAVE EIFFEL
## 1832-1923
## The Eiffel Tower

Man has always wanted to build high, be it for his own satisfaction or as a symbol of power. Trevithick in 1833 proposed a 1000ft high column of cast iron tapering from 100ft diameter at the base to 12ft diameter at the top. For the 1874 Exhibition in Philadelphia a 1000ft high tower was proposed and in 1881 Sébillot returned from the USA with a design for a 1000ft iron lighting tower for Paris.

When it was decided to hold an Exhibition in Paris in 1889 Eiffel presented the idea of a 1000ft tower to the Minister of Commerce and Industry.(At that time the tallest tower built was 570ft high in Turin.) He stated, 'cette ouvrage colossale devait constituer une éclatante manifestation de la puissance industrielle de notre pays, attestant les immenses progrès realisés dans l'art des constructions metalliques'. After seeing his proposals it was decided to hold a competition for a tower of 1000ft height to be built on a square base. Eiffel won it.

Eiffel, born in 1832, studied engineering at the Ecole Centrale des Arts et Manufactures in Paris from 1852-55. He was a year ahead of W.L.B. Jenney. After working on railway construction he set up his own works outside Paris, 'Maison G Eiffel', in 1867. From 1867-69 he designed four railway viaducts in the Auvergne, two of which he built himself. They were of cast and wrought iron latticed trusses and supporting towers very similar to those first used at Crumlin in 1853 (Eiffel refers to Crumlin in a lecture in 1888). The viaduct at Rouzat in 1868  embodies all of Eiffel's mastery of structural elements and connections. He widened the towers at their base to a parabolic form; a shape he developed for his Tower. He summed up his experience thus:

*La rigidité des piles ainsi constituées est très grande, leur entretien très facile et leur ensemble a un réel caractère de force et d'élégance.*

His choice of latticing came from a study of the wind effects on structures, and he used a minimum

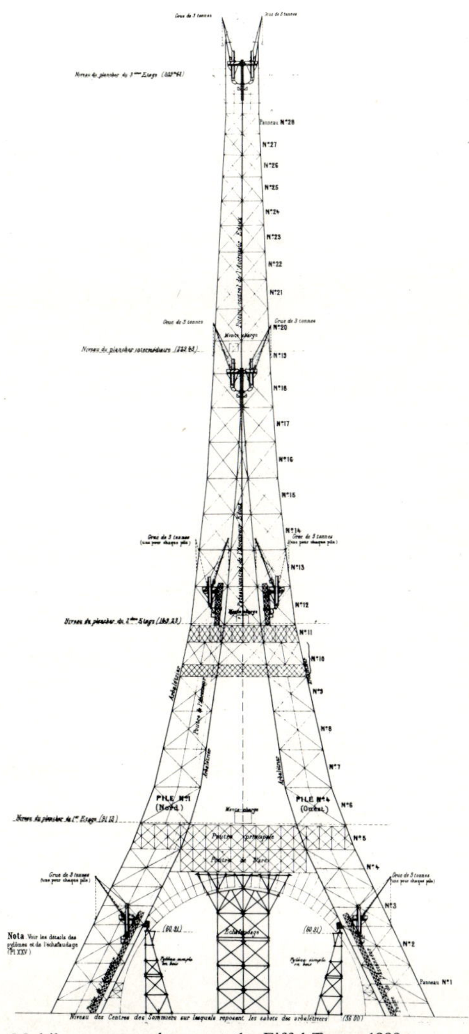

Mobile cranes used to erect the Eiffel Tower 1889.

amount of material exposed to the wind. His interest in the pressure and distribution of wind on structures never left him.

Eiffel followed with his well-known two-pinned parabolic arched bridges in wrought iron, the Douro bridge at Oporto in 1878 and the Garabit viaduct in 1884. The latticed towers were truncated for maximum economy and not curved as at Rouzat, and the latticed tubular arch not only varied in depth but also increased in width from the centre to the springing to form a smooth curve, again to provide maximum stiffness for lateral wind loading and for economy.

His interest in latticed towers was shared by two of his engineers, N.M. Nouguier and Koechlin, who made proposals for a 1000ft tower in about 1885. Eiffel agreed to more studies being carried out and an architect, M. Sauvestre assisted. They concluded that the economic form of the Tower was such that at any level the tangents to the curve of the legs should intersect at the centre of gravity of wind loads above the level considered.

After he won the competition for the tower Eiffel put all his energy into its design and it was said that he employed forty draughtsmen and calculators who worked for over two years on the plans which included 3000 working drawings and 700 diagrams.

He assumed that problems of construction would be overcome. In order to maintain the legs of the Tower in their correct position to allow easy fitting of the fabricated wrought iron bits, as well as equalising the loads, he set two of them on jacks — an elegant solution. His ingenuity can again be seen in his method of erection of the Tower. Above the first level he used the curving lift shafts as guides for his climbing cranes.

Although Eiffel is best known for the Tower he also designed and erected many buildings of which the most remarkable is the Cupola of Nice which he designed with the architect Charles Garnier. He floated the latticed-iron dome in a hydraulic tank so that it could easily be rotated by hand — again a simple and elegant solution.

Of a different nature was the design and fabrication in 1883-86 of the skeleton structure of the Statue of Liberty some 151ft high. Eiffel produced a latticed wrought-iron frame primarily to resist wind loading. At the same time Jenney was solving the same problem in the design of the first skyscraper in Chicago.

As with Freyssinet Eiffel was an engineer's engineer immersed in his work. His great interest in wind loading and aerodynamics led him to retire completely from construction in 1889 at the age of 57 and to devote all his time to the study of aeronautics and to his wind tunnel. He made significant contributions to the design of aircraft and is credited with the idea of the monoplane.

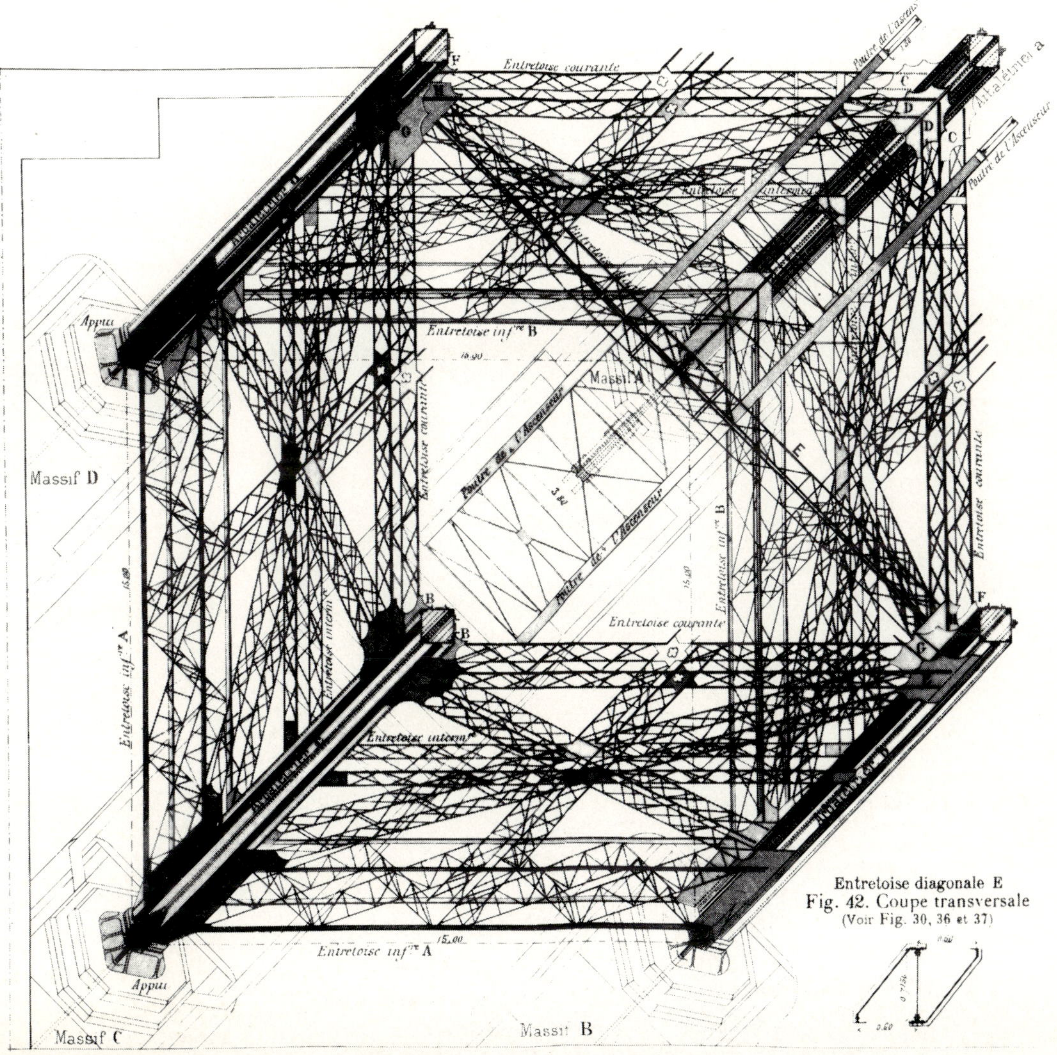


Detail of Ironwork, Eiffel Tower.

## 26  EIFFEL · EXHIBITS

E1   Gustave Eiffel, 'Travaux scientifiques éxécutes à la tour de trois cents mètres', 1900.
Etched portrait frontis.

E2   Nouguier and Koechlin. The original design of 1884 (photograph, ETH, Zurich).

E3   La Tour Eiffel, 1889. Commemorative photograph album.

E4   Eiffel, 'Notice sur le Pont du Douro, à Porto', 1879. Plate 1, elevation of the bridge.

E5   Eiffel, '… le Pont du Douro', 1879. Plate V, erection of the arch.

E6   Eiffel, 'Notice sur le Viaduc de Garabit', 1888. Plate III, construction of the arch.

E7   Henry N. Maynard, 'Handbook to the Crumlin Viaduct, Monmouthshire', 1862.
Decorative commemorative binding and plate showing details of the bridge.

E8   Eiffel, 'La Tour de Trois Cents Metres', 1900. The construction of a foundation.

E9   Eiffel, 'La Tour…', 1900. The successful competition design.

E10  Eiffel, 'La Tour…', 1900. Plates showing details of the ironwork and its construction.

E11  Eiffel, 'La Tour…', 1900. 3 photogravure plates showing progress of erection.

E12  Eiffel and Charles Garnier, 'Observatoire de Nice. Coupole du Grand Equatorial', 1885.

E13  Eiffel, 'Les Grandes Constructions Métalliques', 1888.

E14  Sheet of commemorative writing paper and envelope, 1889.

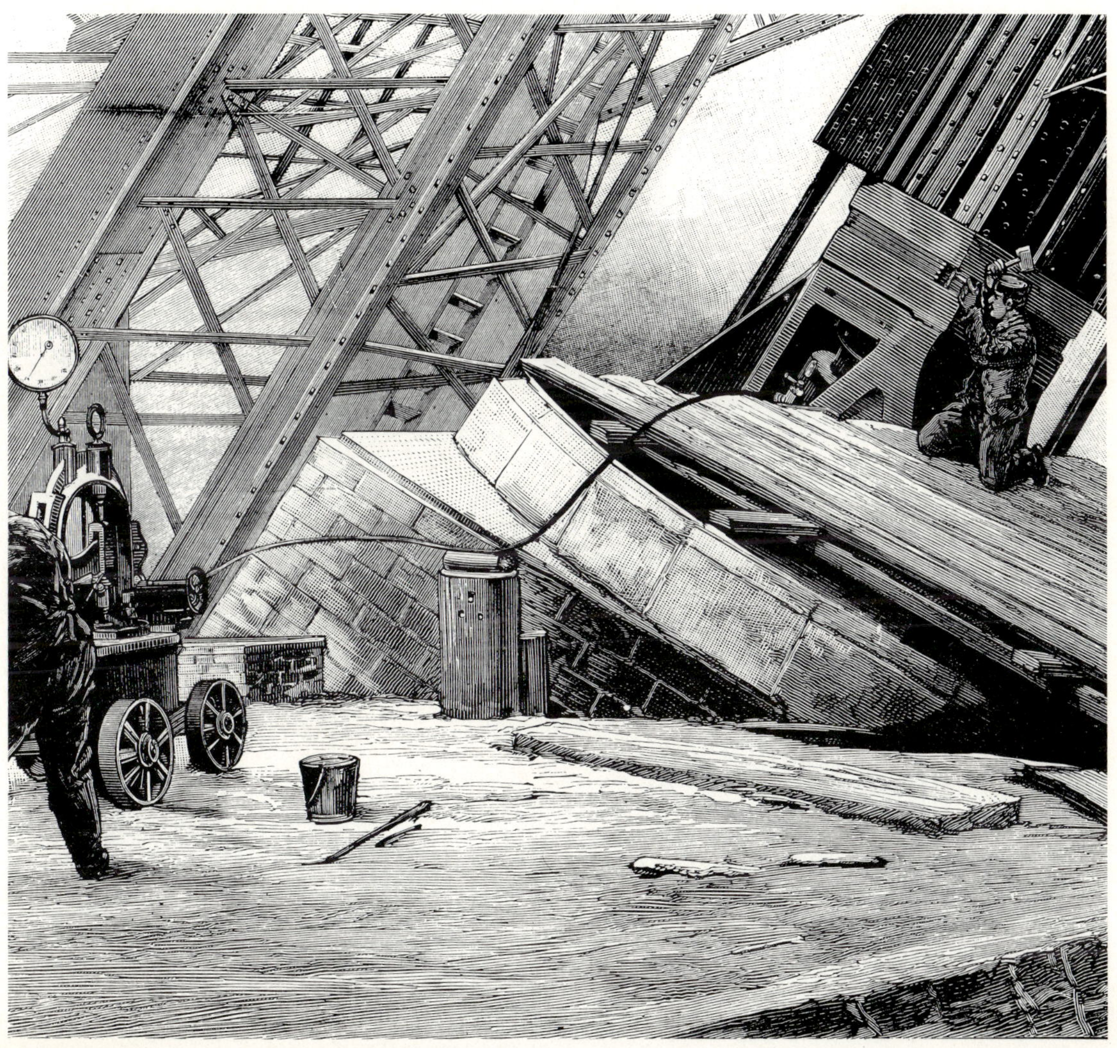



Jacking one of the legs.

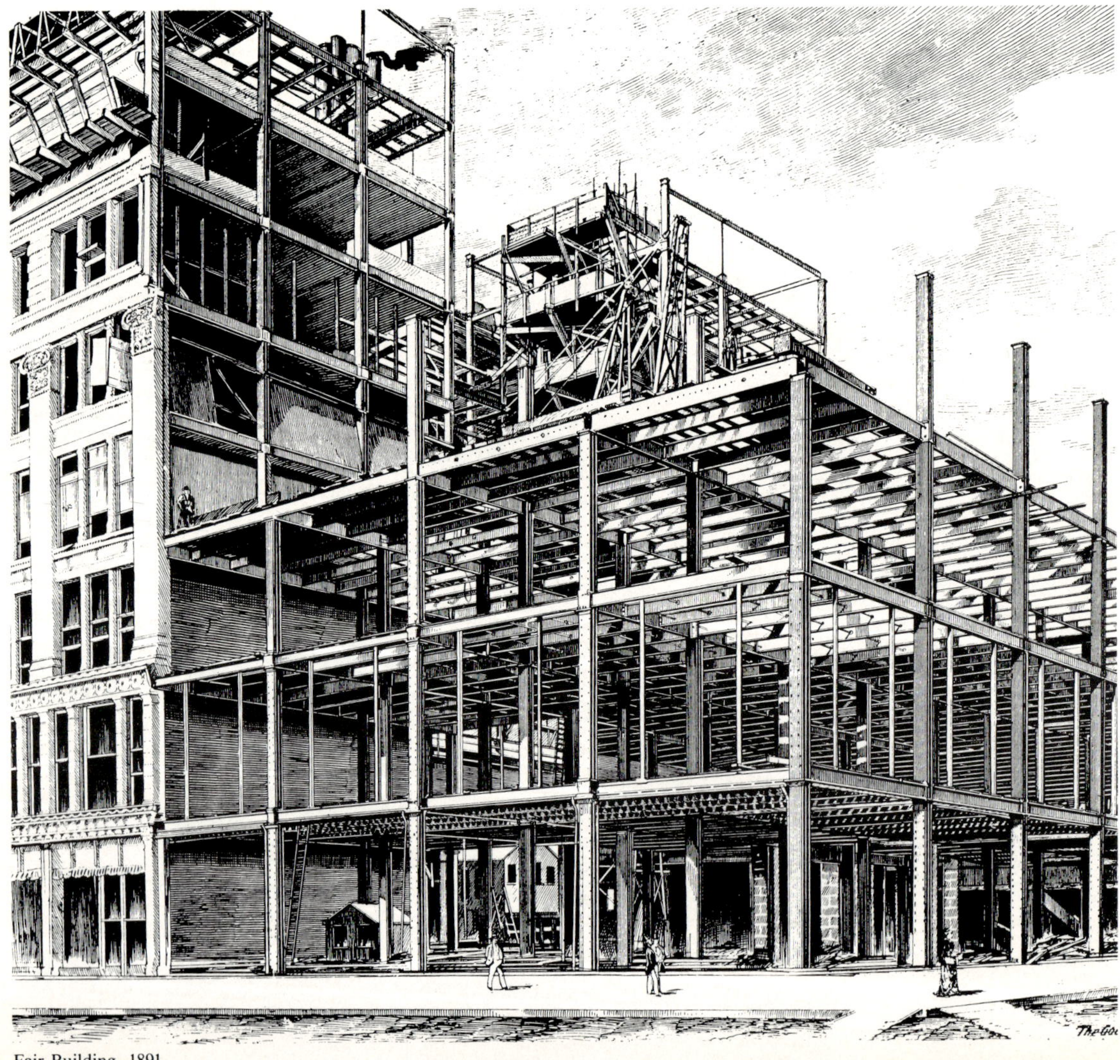

Fair Building, 1891.

# W. LE BARON JENNEY
**1832-1907**
**Home Insurance Building Chicago**

'The Skyscraper is the most distinctively American thing in the world'. So Col. Starret opens his book on skyscrapers in 1928, defining a skyscraper as a tall building with skeleton frame and illustrating the birth of such buildings in Chicago after the fire of 1871. The introduction of the elevator meant that buildings over six storeys were viable and in a period of commercial boom prestigious offices were required. Coupled with increasing land values and rising population in Chicago this posed a new and exciting problem for both architect and engineer.

The early rebuilding followed traditional methods with load-bearing external brick walls and internal cast-iron columns and wrought-iron beams, which were fireproofed in many interesting ways. Lateral stiffness was provided by the brickwork.

The ground conditions in the Chicago Loop were known from borings. Early buildings had foundations taken 15 feet through silt to a layer of stiff clay or hard pan. Loads on the ground varied and, due to an appreciable depth of soft clay below the hard pan, differential settlement was common. As the loads increased with height, settlements became higher but were still accepted. In 1873, Baumann published his method of isolated piers in which he recommended that foundation should be sized so as to apply an equal pressure on the ground and thus reduce differential settlement. This is probably the first scientific design of foundations.

Not all foundation designs followed Baumann's method. Some foundations consisted of a thick mass-concrete bed on the clay over the whole area of the building on which load-bearing walls and columns sat irrespective of their loads. In the case of the Federal Building of 1879 the result was cracking of the concrete and a settlement of 14″ within six years. For the City Hall in 1882 there was differential settlement of 7″ during construction and 14″ before its demolition.

During this period a young architect L.S. Buffington, inspired by the ideas of Viollet-Le-Duc, began to imagine buildings of twenty to one hundred storeys high with an iron framework and in 1880 produced outstanding designs which, in their way, are relevant today. He called them 'cloud scrapers'. He carried out calculations and patented his ideas in 1888. However it was W.L.B. Jenney who in 1884 built what is considered to be the first skyscraper: the Home Insurance Building.

William Le Baron Jenney, architect and engineer, was born in 1832 in Massachusetts and studied engineering at Ecole Centrale des Arts and Manufactures, Paris, between 1853 and 1856. After service in the army he set up an office in Chicago in 1867 and young architects such as Sullivan, Holabird, Roche and Burnham all worked there. 'William Le Baron Jenney', Giedion wrote, 'played much the same role in the training of the younger generation of Chicago architects that Peter Behrens did in Germany around 1910, or Auguste Perret in France'.

In 1883 he was appointed as architect for a proposed tall fireproofed building. The President of the Home Insurance Company wrote to Mr. Jenney that he desired the maximum of well lighted offices above the second floor, knowing that there would be insufficient external brickwork to carry the loads at the lower levels. Jenney appreciated that iron columns would have to be introduced into the brick piers and had to solve the problems of relative temperature movement between the iron and brick over a height of 150 feet. He resolved to break the brick facade at each floor level and to carry its load on the iron columns. He described his proposal as a simple, straight-forward engineering problem, resembling railway bridges standing on end, and would put his designs and calculations before any good bridge engineer and abide by his decision. His design was accepted and the first skeleton frame was built.

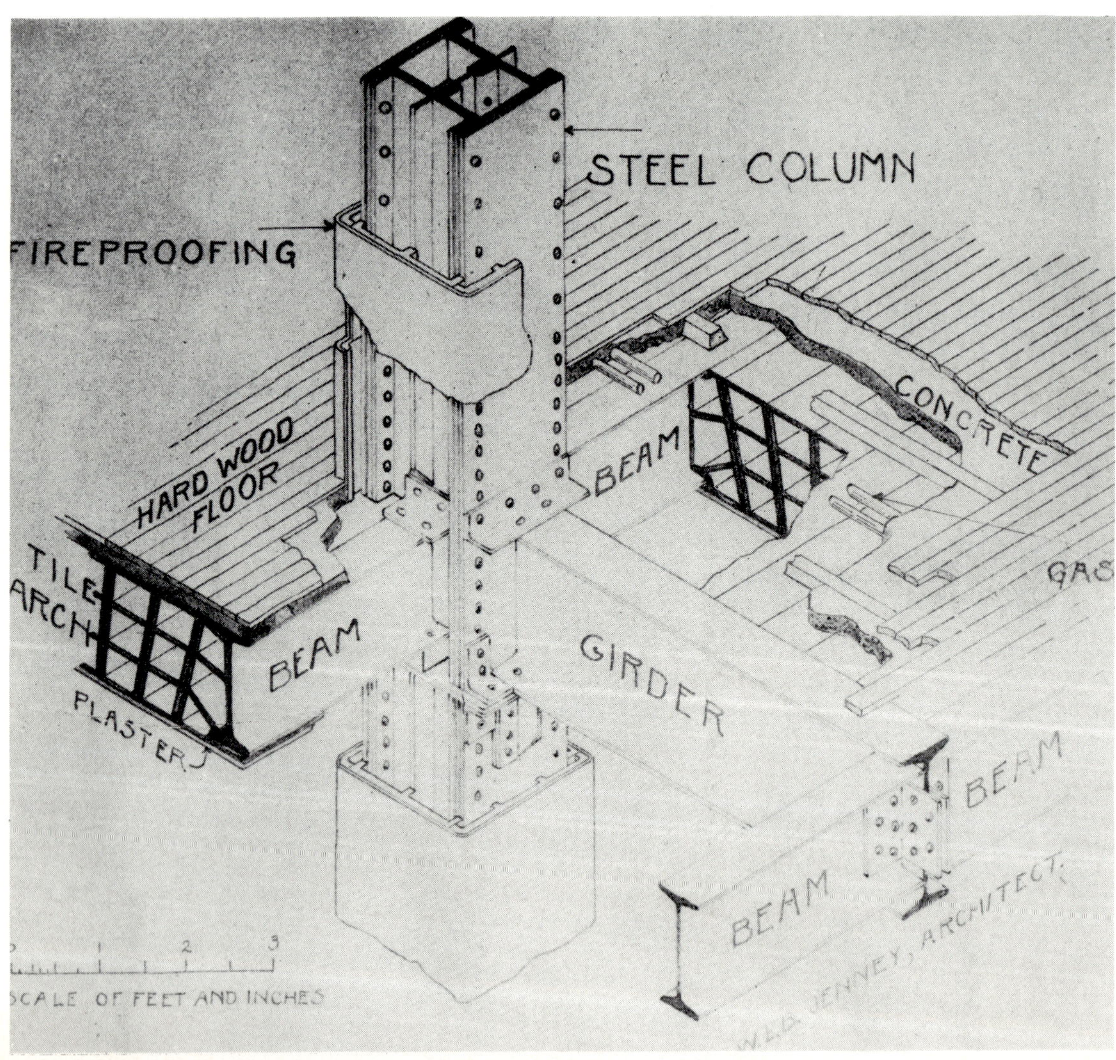


View of an interior column, Fair Building 1891.

Foundations followed Baumann's method of independent piers. The building was set 2½″ high to allow for settlement and at the end of the first year the relative settlement was in the order of ¾″. A foundation pressure of 2 tons per square foot was used, based on the results of twenty borings on site.

Jenney's system of construction was appreciated by many and the age of the skyscraper had begun. The performance of tall buildings was monitored and reflections and periods of vibration were measured under extreme wind loads as early as 1894. The introduction of the Chicago Caisson method of foundation in 1899, whereby foundation loads are transferred to the hard clay or rock, gave impetus to the increase in height of buildings which by 1925 had reached 45 storeys and today are over 100 storeys.

## JENNEY · EXHIBITS

F1    William Le Baron Jenney. Portrait from 'Industrial Chicago, The Building Interests', 1891.

F2    'Industrial Chicago', 1891.

F3    Frank A. Randall, 'History of the Development of Building Construction in Chicago', 1949.

F4    Carl W. Condit, 'The Chicago School of Architecture', 1964.

F5    Home Insurance Building. Diagram of Iron Work — Exterior Walls — 2nd storey. Photograph.

F6    The Chicago Sunday Tribune, June 23, 1907. Front page.

F7    Copy of a letter from Eiffel to Jenney, 1889.

F8    Leroy S. Buffington. A group of drawings from his patent for iron building construction, 1888.

F9    Leroy S. Buffington. A group of drawings showing his 'Cloudscraper' project, 1888.

F10    Col. W.A. Starrett, 'Skyscrapers and the men who built them', 1928.

F11    'Proceedings of the International Conference on Planning and Design of Tall Buildings', 1972.

F12    A lift-gate designed by Louis Sullivan for the Chicago Stock Exchange.
(B. Weinreb Architectural Books Ltd.).

# MAILLART & FREYSSINET



From the earliest days engineers have searched for a material to connect together stones and bricks, particularly below water. The Romans discovered pozzolano, a natural cement, which they used for their magnificent concrete structures such as the dome of the Pantheon and which, much later, was imported by Smeaton for his tests on hydraulic mortars during the building of the Eddystone lighthouse. Other natural cements appeared in the early 19th century and the first man-made cement was invented and its manufacture patented in 1824 by Aspdin who called it 'Portland Cement'.

Much use was made of it for fireproofing floors and in time experiments showed that the tensile strength of the material could be substantially increased by the addition of steel reinforcement.

Patents for reinforced concrete floors were taken out by Wilkinson in England and in France by Coignet, who had used concrete in place of stone, and by Monier for reinforced concrete tanks and pipes.

In 1880 Hennebique, a building contractor, was so impressed by the potential of reinforced concrete that he privately carried out tests on floors and load-bearing frames which were successful. His system, patented in 1882, heralded a revolution in the building industry and architects began to design with this new material in mind.

The Paris Exhibition of 1900 made the first extensive use of reinforced concrete in many of its pavilions and ground works. The various systems were illustrated in reports on the exhibition and Coignet's son and Hennebique were each awarded a Gold Medal for their pioneering work.

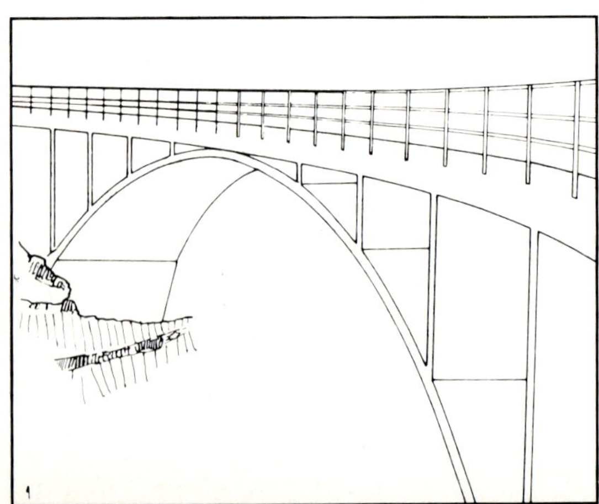

Schwandbach Bridge 1, 1933.

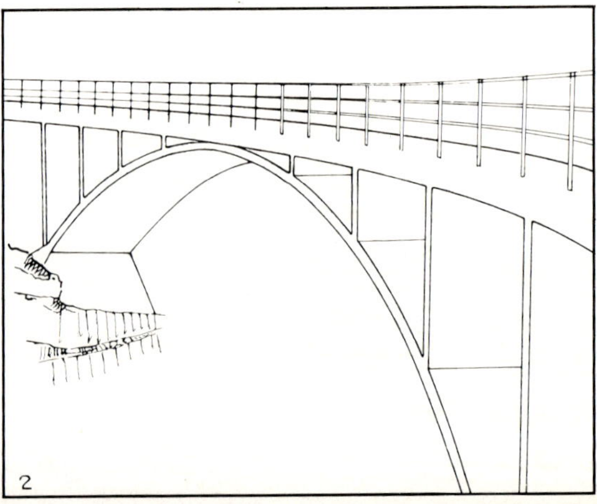

Schwandbach Bridge 2, 1933.

A Committee was formed in 1900 in Paris to investigate test data and to propose, as a safeguard to society, a set of regulations for the use of this new material; subsequently these were published in Paris in 1906.

Christophe in 1902 published his book *Le Béton Armé et ses Applications* which gives a good view of the state of the art at the time when two engineers, Maillart and Freyssinet, were starting on their illustrious careers.

## MAILLART & FREYSSINET · EXHIBITS

G1  'Hennebique's patent system of ferro-concrete construction', 1905 (The Science Museum).

G2  'Reinforced concrete constructions. Coignet system', c.1910 (The Science Museum).

G3  Coignet and Tedésco. 'Du calcul des ouvrages en ciment avec ossature métallique', 1894.

G4  'Indented steel bars for reinforced concrete construction', 1909 (The Science Museum).

G5  'Kahn system illustrated', 1921 (The Science Museum).

G6  Kahn trussed and rib reinforcing bars.

G7  'Revue technique de l'Exposition Universelle de 1900. Architecture et construction', 1900, 1901.

G8  Paul Christophe. 'Le Béton armé et ses applications', 1902.

G9  Thaddeus Hyatt. 'An account of some experiments with Portland-Cement-Concrete combined with iron, as a building material...', 1877.

G10  Charles F. Marsh. 'Reinforced concrete', 1904.

G11  'Le ciment-roi. Réalisations architecturales récentes', nd.

G12  Project for a presidential palace c.1910 by A. de Baudot. Photograph from Peter Collins, 'Concrete', 1959.



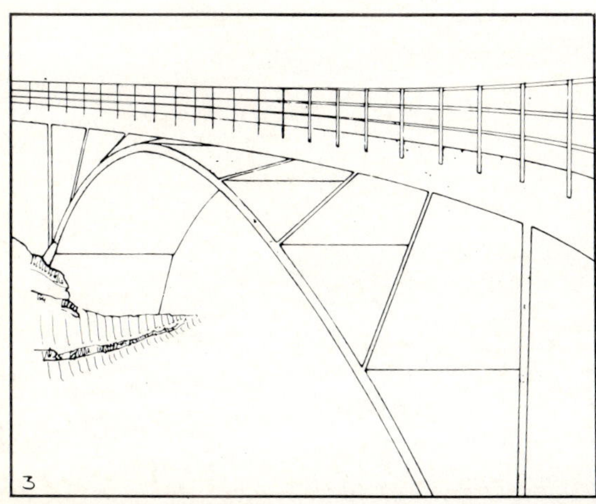

Schwandbach Bridge 3, 1933.

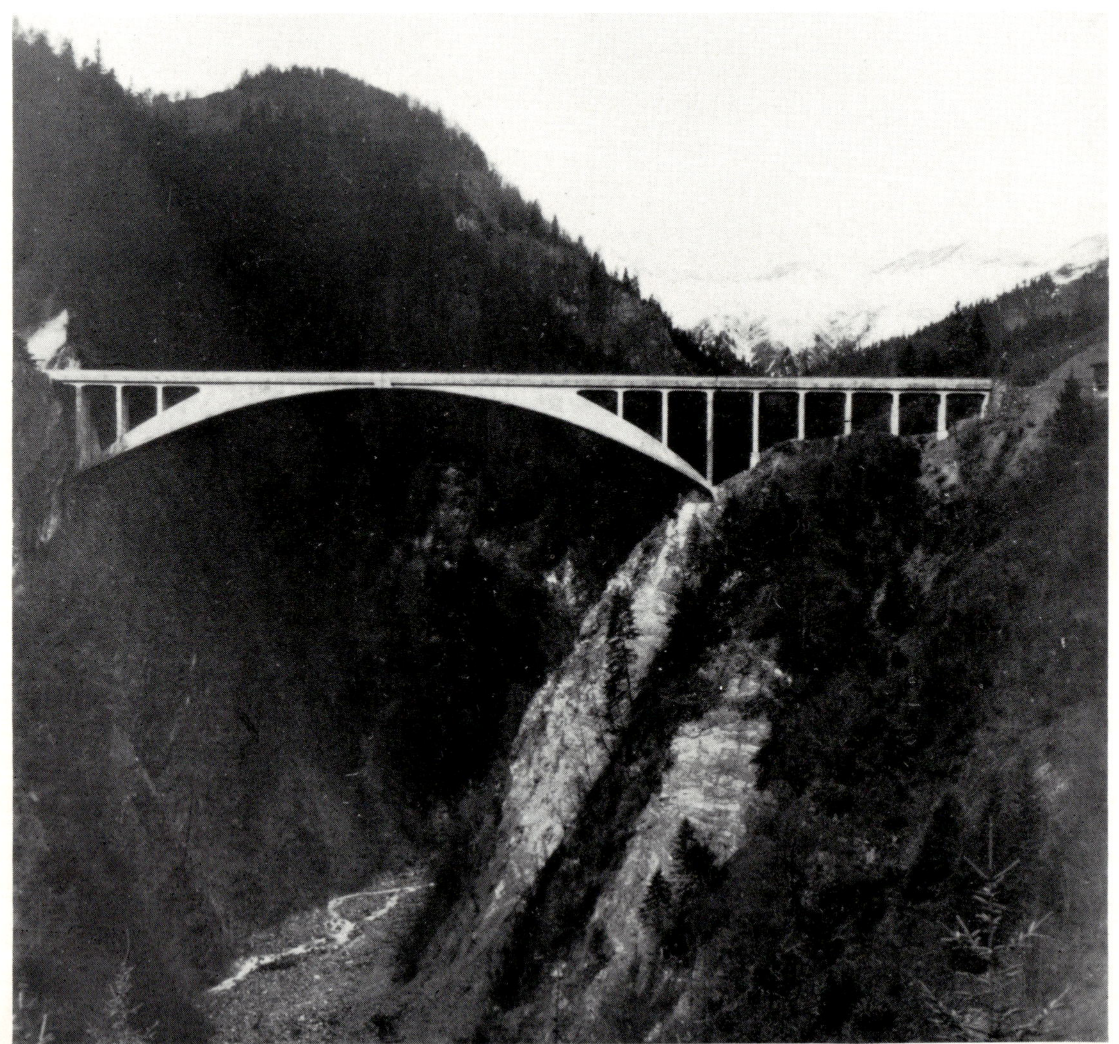

Salginatobel Bridge, 1930

# ROBERT MAILLART
**1872-1940**



## THE TAVANASA BRIDGE 1905

In 1905, some thirteen years after Hennebique's patent, Maillart designed and built the 170ft span arched Tavanasa bridge of 19ft rise in which he exhibited his emerging mastery of reinforced concrete and his flair as an engineer.

In mountainous terrain, the art of bridge building has always been highly developed. Elaborate long-span timber bridges appeared in the mid-18th century through the work of the Grubenmann brothers in Switzerland. The craftsmen were skilled and the timber plentiful and each local community built its own bridges. In the early 1900s the birth of reinforced concrete meant that local labour and materials could again be used for bridge building instead of importing expensive wrought iron and steelwork. Acceptance of the new material was quickly forthcoming.

Robert Maillart, born in 1872, graduated from ETH Zurich in 1894. Although a Monier system arched bridge had been built in 1890 near Zurich there had been no course on reinforced concrete at the University. His professor, Ritter, was apprehensive of the working together of steel and concrete under temperature change and vibration. In 1899 Maillart built his first unreinforced or mass concrete bridge. In 1900 he went to the Paris Exhibition to study the various systems of reinforced concrete, and the following year, after attending a lecture by Hennebique's agent in Switzerland, he wrote in *Schweizerische Bauzeitung* that he considered reinforced concrete to be the material of the future.

His knowledge of the material was increased by the construction, in 1901, of the Zuoz bridge. It was of similar form to that of his first bridge but for economy Maillart replaced the solid arch with a thin arch stiffened by spandrel walls. Maillart was looking for economy not only in materials but in formwork. He built the thin arch first and then used it as centering for the remainder of the structure.

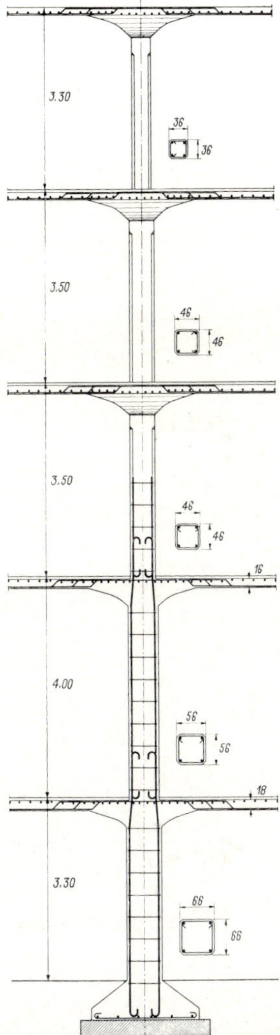

Maillart's flat slab system, 1909.

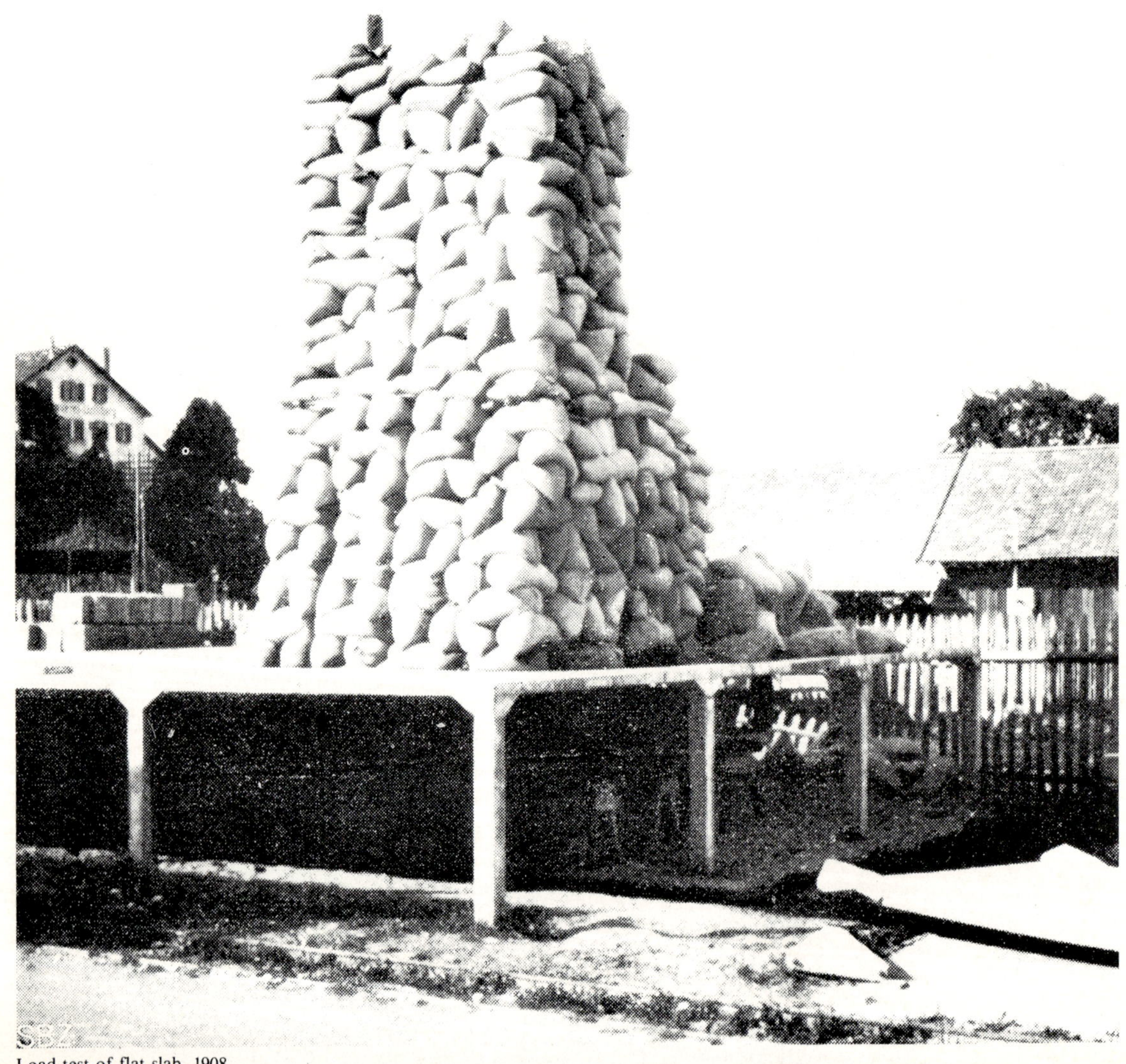

Load test of flat slab, 1908.

In time, minor cracking appeared in areas of the spandrel walls of the Zuoz bridge. Maillart appreciated that their stiffness was attracting load for which they were not reinforced and in his Tavanasa bridge he corrected his error. Its form, with hinges at the centre and at the abutments, approached structural purity and later became a Maillart hallmark. He had then been working on his own as a designer and contractor for only three years. As early as 1900 Maillart had stated his design philosophy with regard to bridges. He considered that their efficiency should be proved by test loading, that they should be of minimum cost but also be of elegance visually.

The Tavanasa bridge fulfills all these ideals.

## KLOSTERS RAILWAY BRIDGE 1930

In 1908 Maillart started experiments on concrete flat slabs which he considered a rational use of the material. The load tests were successful and the next year he was granted a patent. His system was far superior to the only other proposal by Turner in the USA. Maillart's Warehouse in Zurich built in 1910 with flat slabs was one of the first of its kind and again had Maillart's stamp. After being marooned in Russia during the war he returned to Switzerland to practise solely as a consultant.

Observation of cracks in a parapet beam of an arched bridge led him to appreciate that when the arch was loaded on one side only the stiffness of the parapet restrained the arch from fully deflecting. With a relatively stiff reinforced parapet he could reduce the arch thickness to the minimum. This he did in his 1923 Flienglibach bridge which was the first of a new generation of concrete arched bridges. For the Klosters bridge of 1930 he used the same principle but curved the deck on plan. He combined the inherent stiffness of the arch, the vertical walls, the deck slab and the parapet beams to resist torsion thereby producing a truly three-dimensional form.



In 1938 Maillart wrote, 'Reinforced concrete does not grow like wood, it is not rolled like steel and has no joints as masonry. It is more easily compared with cast iron as a material cast in forms... The engineer should then free himself from the forms dictated by the tradition of the older building materials so that in complete freedom and by conceiving the problem as a whole it would be possible to use the material to its ultimate'.

Max Bill describes Maillart as having 'the creative power of an artist who always conjures up something new through the means of expression of his time and by making use of all the possibilities at his disposal'. It is this flair for design which excites architects and engineers alike.

## MAILLART · EXHIBITS

G13   Robert Maillart.
Photograph from obituary issue. Schweiz Verband fur die Material Prüfungen der Technik, April 1940.

G14   Schweizerische Bauzeitung for May 25, 1901.

G15   Billington. 'Maillart's Bridges'.
3 sectional drawings of Stauffacher bridge, Zuoz bridge and Tavanasa bridge.

G16   Robert Maillart. 'Bogenträger aus armiertem Beton'.
Patent number 25712, class 5, Schweizerische Eidgenossenschaft, 1902, for arched beams.

G17   Tavanasa bridge.
Photographs (ETH, Zurich).

G18   Salginatobel bridge.
A group of 3 photographs showing construction and the completed bridge.

G19   Maillart & Cie, 'Konstruktion zur Raumabdeckung'.
Patent nr 46928, class 4a, Schweizerische Eidgenossenschaft, 1909, for flat slabs.

G20   Klosters railway bridge.
A group of photographs showing the actual bridge with details of the reinforcements and centering. (ETH, Zurich).

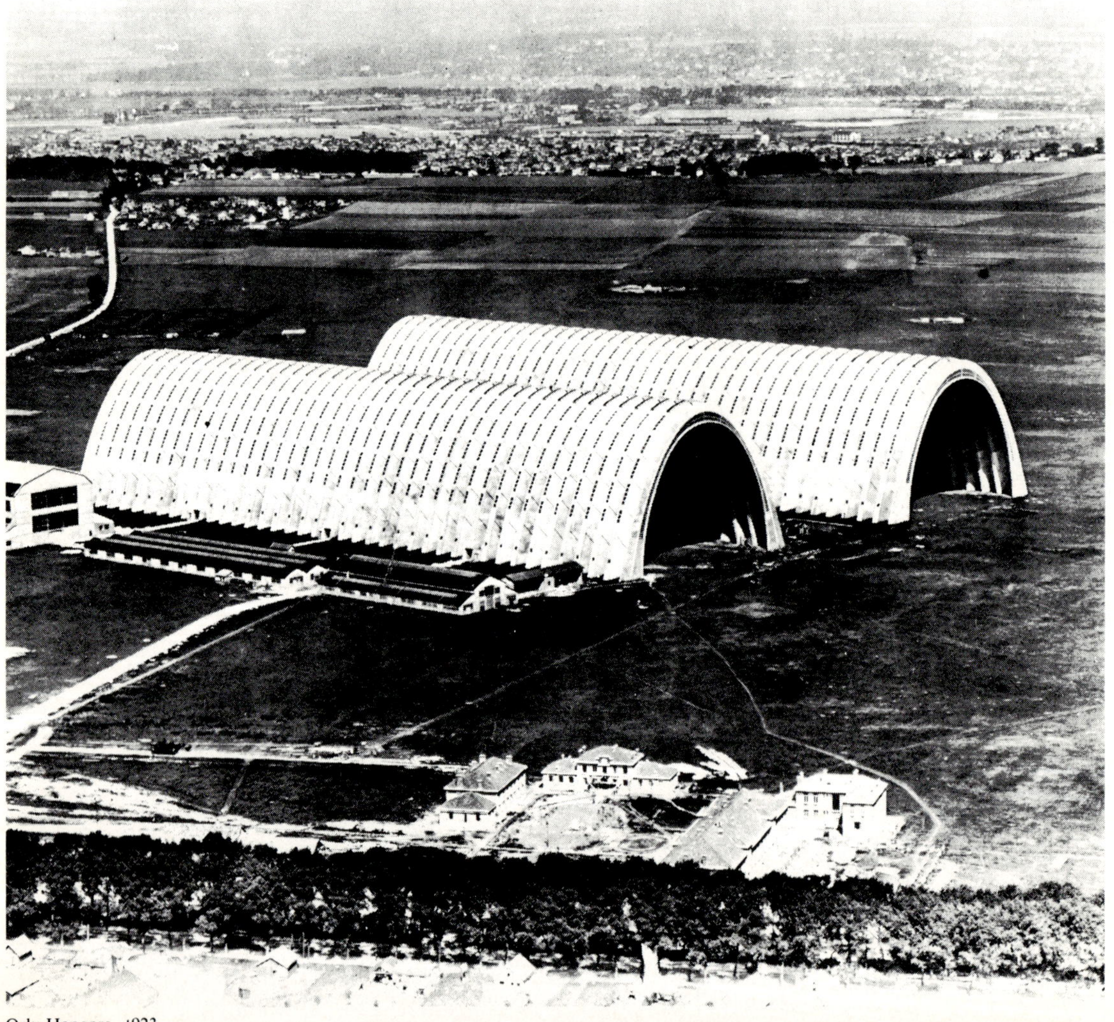

Orly Hangars, 1923.

# EUGENE FREYSSINET
**1879-1962**
**The Veurdre Bridge 1910**

At the time Maillart was completing his Tavanasa bridge Freyssinet was finishing his studies in Paris. The design of reinforced concrete structures was taught by Rabut who, with Hennebique, did not ratify the 1906 Paris regulations.

Freyssinet was appointed Rural Services Engineer for an area near Moulins, assisting the local community in building small bridges and culverts for low cost. For large projects he worked with his director engineer Mercier. In 1907 Freyssinet used jacks for the first time instead of sand boxes to decenter a 110 foot span arched bridge. It was an inventive solution to a traditional problem.

The following year Freyssinet was allocated sufficient funds to test a prototype shallow arch of 230 foot span. As the ground at the test site was not capable of carrying the horizontal thrust from the arch, he tied the abutments together with a concrete member precompressed by high-strength steel strands - the first application of prestressing. To decenter he introduced jacks at the springings and by applying compression he lifted the arch until it was freestanding — a most elegant solution. After a successful test (although the result in no way related to the predictions based on the 1906 regulations), Freyssinet was sufficiently confident to start building the Veurdre bridge of 1910 which had three spans of 225 feet, 240 feet and 225 feet. Decentering jacks were positioned at the crown of each arch which by regulation had to be designed as a three-hinged frame. The bridge was tested with steam rollers but within a year the crowns had deflected over 5 inches. He realised that due to the movement of concrete in time under load (ie creep), the bridge was shortening and dropping. He rejacked the centre and all was well. However his successful test arch, with which no such problems occurred, was two-pinned and he turned to this form for future bridges.

This contrasts with Maillart who recommended three-pin arches so as to reduce thermal stresses but whose arches had a much smaller span to depth ratio and so deflected less.

Freyssinet spent a great deal of time exploring the properties of the material particularly creep and experimenting with mixes to produce high strength concrete. He wrote on conception of structure:

'There are two possible sources of obtaining information: one is a direct perception of reality, the other is intuition. By this I mean the sum total of human experience built up over the years in the subconscious.'

'It is only natural that intuition should be controlled in the light of experience, but when it turns out to be in direct opposition to some calculated result I always make a double check on the calculation... in the long run, it is almost always the answer determined by calculations that turns out to be wrong.'

## PATENT ON PRESTRESSED CONCRETE 1928

In 1918 Freyssinet joined a contractor to design and build both bridges and buildings. His versatility is shown in his construction of reinforced concrete boats from 1918-21 and in the now famous hangars at Orly made from a 8cm thick corrugated shell, 190ft high and some 295ft in span. He later used shell roofs for factories and for the Austerlitz Station in Paris.

Freyssinet's main interest however was in prestressing. After patenting his system he left his firm, Société Limousin, on 1st January 1929 at the age of 50 to concentrate solely on prestressing.

'By prestressed construction, I understand a structure which is subjected to a system of artificially produced permanent forces before the application of the imposed loads, or in the case of permanent loads, at the same time. These supplementary stresses thus created, preferably in the opposite direction to those due to the loads, are of such a value that the resultant tensions from the combination of applied forces do

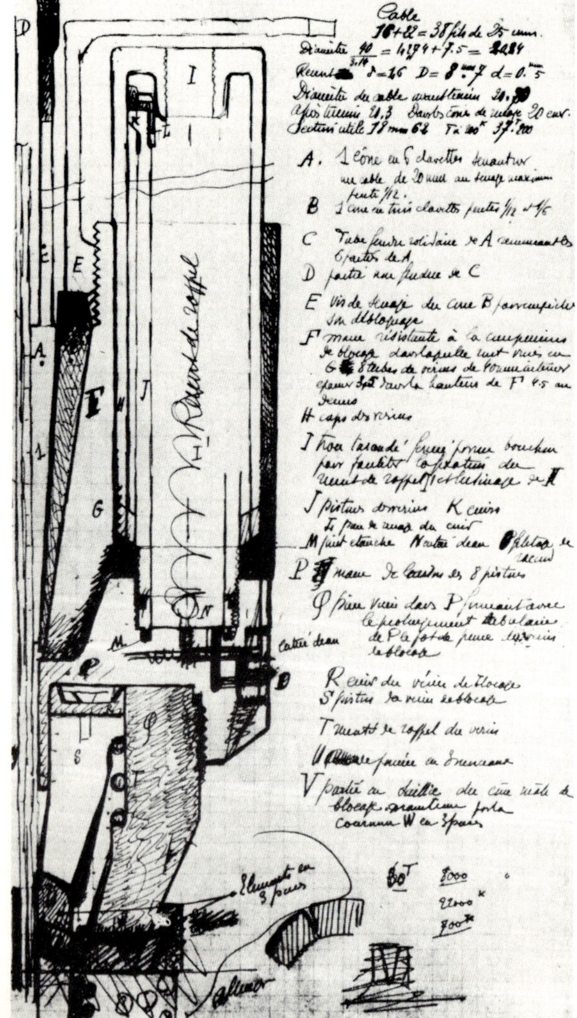

Design for prestressing jack, 1939.

not bring about any stresses which the materials cannot bear indefinitely and with complete safety.'

'A system of prestressing is not limited to a predetermined technical field. It is, in reality, a state of mind, an affirmation of the will of the engineer to accept no longer the consequences of the initial elastic states brought about at random by the method of construction. On the contrary, using prestressing as one more datum of his projects, the engineer can modify this state at will, just as if dealing with the resistance of the beams.'

By 1931 he was producing prestressed concrete lamp-posts in a special factory having solved all the technical problems of high-strength concrete but there was no market and Freyssinet lost a fortune.

In 1934 his luck changed when at the last minute he saved the Marine Terminal at Le Havre from collapse (it was settling at the rate of 1 inch per month) by adding extra concrete between existing foundations and by prestressing them together to form a continuous grillage. Prestressed concrete piles were introduced as a permanent foundation. From that time onwards the idea of prestressing was accepted and in 1936 his first bridge, spanning 64ft, was built. In 1939 Freyssinet patented his jack and conical anchorage which are still in use today.

Freyssinet, unlike Maillart, was an engineer's engineer. He spent all his time seeking for truth in the behaviour of materials and structures, lacking interest in art and architecture. However his structures and his monumental conception of prestressing have provided the greatest inspiration to twentieth century engineers.



G21   Eugène Freyssinet.
Photograph from Ordonez 'Eugène Freyssinet', 1979.

G22   Drawing of 1908, trial arch.

G23   Plans, elevation and centering of the Veurdre Bridge.
Photograph from Ordonez 'Eugène Freyssinet'.

G24   Shell roof of the National Radiator factory at Dammarie-Les-Lys.
Photograph from Ordonez, 'Eugène Freyssinet'.

G25   E. Freyssinet, 'Les Hangars à Dirigeables de l'Aéroport d'Orly', 1923.

G26   Two contemporary photographs of the hangars at Orly under construction.
From Jean Badovici, 'Grandes Constructions. Béton armé, acier et verre'.

G27   Drawing and photographs of the marine terminal, Le Havre, showing Freyssinet's remedial work.
From Ordonez, 'Eugène Freyssinet'.

G28   Reinforced concrete boat under construction.
Photographs from Ordonez, 'Eugène Freyssinet'.

G29   Hollow, pre-stressed poles at Montargis factory.
Photograph from Ordonez, 'Eugène Freyssinet'.

G30   Freyssinet's sketch design for a prestressing jack.
From Ordonez, 'Eugène Freyssinet'.

G31   A Freyssinet prestressing jack.
(Chalkpits Museum).

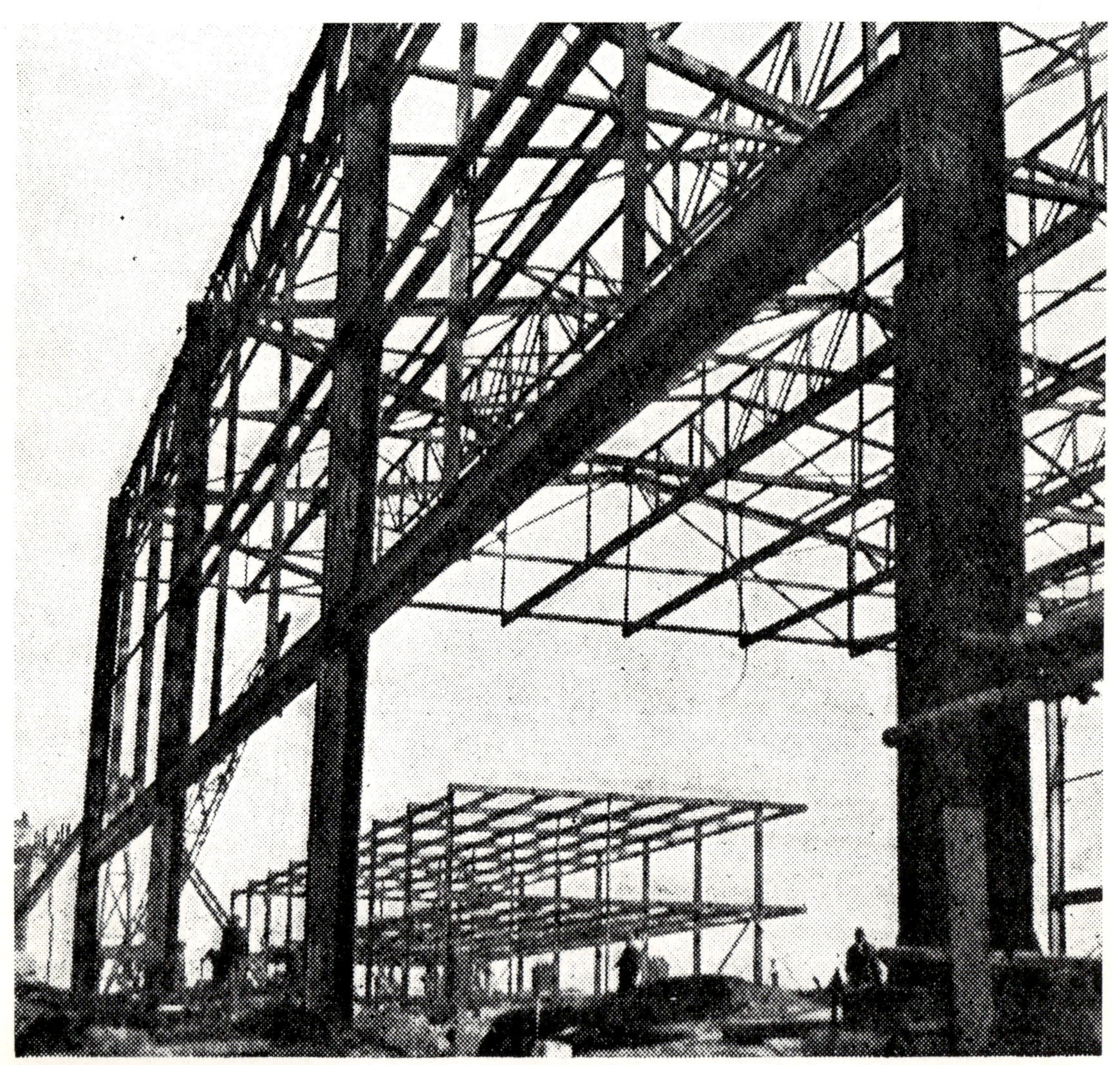

De La Warr Pavilion, Bexhill, 1935.

# FELIX J. SAMUELY
## 1902-1959
### De La Warr Pavillion, Bexhill 1934



The competition for the Pavilion was won by architects Chermayeff and Mendlesohn, who had recently left Germany. The structure was the first all-welded steel construction in England, and was designed by Felix J. Samuely.

Born in Vienna in 1902, Samuely studied in Berlin, where he obtained his engineering degree in 1923 and, apart from a year in an architect's office in Vienna, worked with contractors up to 1929 when he started his own practice with Stephen Berger. This lasted until 1931 when he went to Russia. During those two years Samuely designed the first commercial welded-steel building in Berlin, worked with Erich Mendlesohn and built a factory with Arthur Korn. On welding he writes, 'I have worked on the problem of welding of steel construction since 1928 when I had occasion to design an experimental, welded structure for a water cooler for Siemens Schuckert. This tower was constructed as an imitation of a riveted building and it struck me forcibly that this was the wrong method of procedure and that welded steelwork should have its own design'.

After two years in Russia working on the design of a steelworks and doing research on steel construction Samuely returned to Berlin via China and came to England in 1933. He obtained work with J.L. Kier contractors, his first job being the calculations for the Penguin Pool, and on winning the Bexhill competition Mendlesohn and Chermayeff invited him to act as engineer. He set up practice with Cyril Helsby to be joined later by Conrad Hamman.

In 1934 Samuely had written a book on welding which was updated in 1940 but never published. It remains in draft form.

In describing the structure of the Pavilion in 'The Welder' in April 1935 the partners wrote, 'During the last two years remarkable advances have been made in Great Britain in the welding of structural steelwork and also in an allied constructional industry, namely, shipbuilding... Several British Standard Specifications related to different applications of the art have already appeared... Finally, in the decision of the London County Council to permit the erection of welded buildings... one of the most important steps has been taken.'

The structure was made up from welded trusses, box columns and simple rolled sections. Joints were positioned in the structure at points of minimum stress so as to ease site welding. Although the structure was not exposed or expressed as welded steelwork it provided stimulus to the building industry.

Samuely followed this with his welded latticed steel trusses made up from tubular chords and rod diagonals, the prototype for the lightweight steel trusses used particularly in school construction after the 1939-45 war. He also used welded pressed metal sections and for Simpsons of Piccadilly proposed a welded Vierendeel truss, one storey high, as a support to external columns; this was too revolutionary for the L.C.C. to accept. The practice prospered and Samuely was the engineer for many of the Modern Movement buildings of the late 30s. He collaborated with such architects as Wells Coates, Emberton, Goodhart-Rendel, Pilichowski, Lasdun and Connell, Ward & Lucas on steel and concrete structures.

He was also an active member of the MARS group and, with Arthur Korn, was joint author in 1942 of the MARS Plan of London. He had a passionate interest in transport and had proposed his own rail plan for London.

His work with architects led him naturally to the Architectural Association where he became the principal structures lecturer during the war, and carried on into the early postwar years. His lectures were memorable and fortunately were recorded.

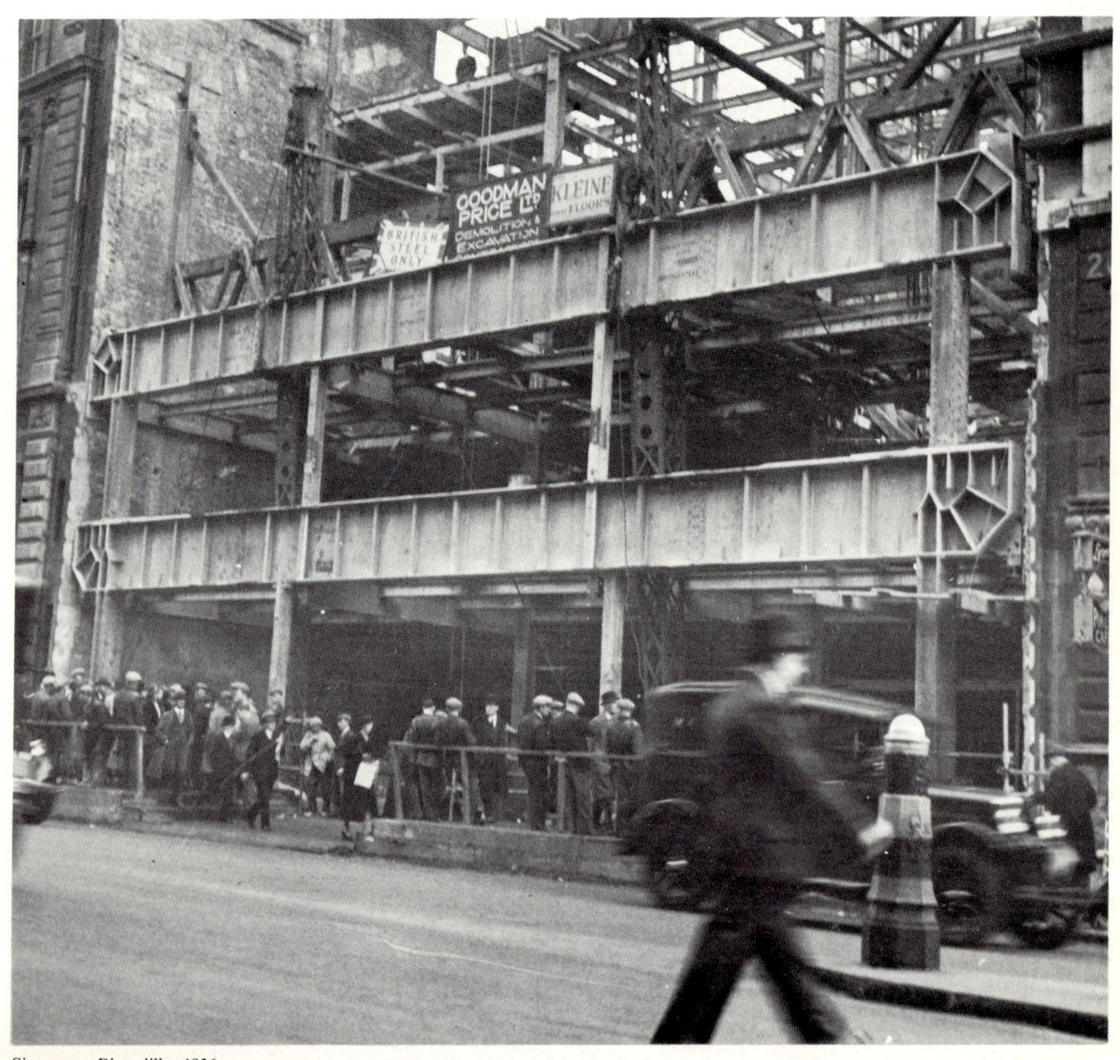


Simpsons, Piccadilly, 1936.

# FACTORY AT MALAGO, BRISTOL 1949

With the wider availability of high tensile wires in 1945, prestressed concrete became a viable structural material. A shortage of timber for shuttering and of skilled craftsmen led Samuely, now practising on his own in his first major postwar construction, to design his most inventive structure of precast and prestressed concrete. The heavily loaded floors were supported on three-hinged precast concrete frames with prestressed concrete ties.

His design philosophy for joints between precast concrete units is readily seen. He treated precast concrete as steelwork and had mechanical connections capable of carrying construction loads and taking up manufacturing and erection tolerances. The floors themselves were of composite construction, of thin coffered precast concrete units with a structural topping of poured concrete which also tied together all the precast concrete units, giving overall stability to the building. For long-span floors he used factory-

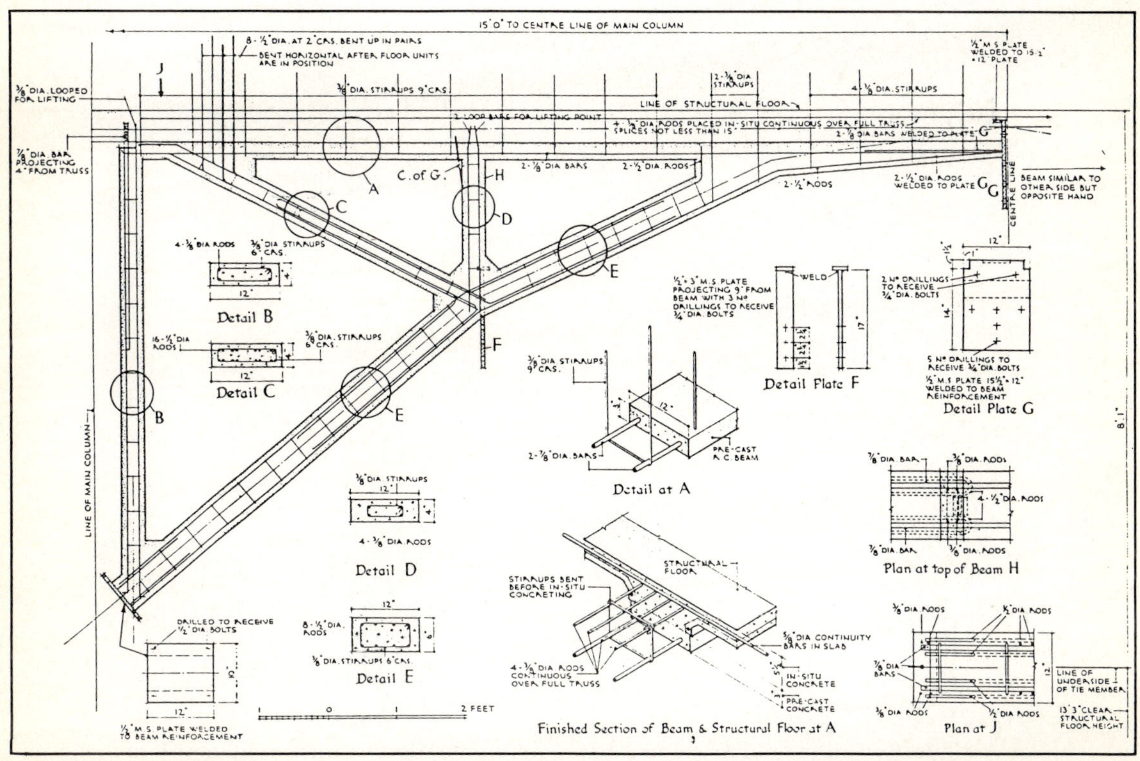

Malago Factory, Bristol, 1949.

46 made prestressed concrete planks in a similar way to steel reinforcement, placing them at the bottom of the slab at mid-span and at the top over the internal supports. The continuous foundation beam, which transmitted the load from the columns into the ground, was also of prestressed concrete using the post-tensioned Freyssinet system. Instead of curving the cables, he curved the concrete beams, which were out of sight below the ground, and used straight cables — a most ingenious solution.

Composite construction was to become his standard method. He made many tests to confirm his ideas and for major types of construction carried out full-scale loading tests.

## SALTASH BAKERY 1948

As early as 1937 Samuely proposed a sawtooth roof for the factory for Steel Ceilings Ltd., using in one plane a welded Vierendeel girder and in the other corrugated steel roofing spanning the apex and the valley and braced by diagonal ties to act as an inclined girder. The two acted together as a folded plate so that no internal ties were necessary except at the columns. This was his first proposal for a three-dimensional roof in England. A more elaborate proposal for an aircraft hanger for BEA appeared in 1947 and a year later he actually built his first folded plate roof of in situ concrete construction.

Addressing the RIBA in 1952, Samuely spoke of space structure, 'I think that, at the moment, we are on the eve of a great revolution, and that hundreds of years hence, people will look back on this time as being the one when construction changed over from 'plane' to 'space' and saw the birth of a new architecture'. Certainly he carried on developing folded plates from concrete, steel and timber, for in them he saw greater potential for both economy and architectural expression than in curved shells.

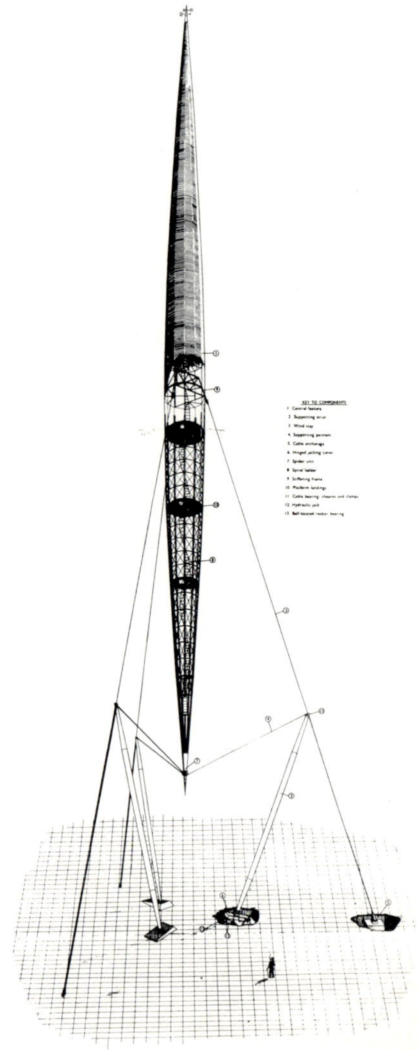



'Skylon' Festival of Britain, 1951

# THE SKYLON, FESTIVAL OF BRITAIN 1951

Freyssinet referred to prestressing as a state of mind and in the construction of the Skylon Samuely further showed his ingenuity. Originally the architects proposed a cigar-like feature suspended in mid-air from inclined pylons. Samuely appreciated that heavy cables would be required to control movement and vibration under wind conditions unless the cables were pretensioned, and so capable of carrying compression. This he achieved by jacking up the pylons in a manner reminiscent of Eiffel. Pretensioning the cable also meant that two of the three proposed bracing cables to each pylon could be omitted.

Samuely did not limit himself to prestressing steel and concrete. In 1952 he prestressed slender brick piers (in order for them to carry wind loads), timber trussed beams and a timber folded plate roof. He also combined materials for maximum economy and efficiency using, for example, prestressed concrete gutters as the tension member of a latticed-steel folded plate roof.

In 1960, shortly after Samuely's death, the Architectural Association devoted the whole of the June issue of its Journal to his life and works. Not only was he a creative engineer and brilliant theoretician, but he fervently believed in architect-engineer collaboration. He will be remembered for his great influence on architects in a period when the conception of structure was changing from two to three dimensions.

## SAMUELY · EXHIBITS

H1   Photograph of F.J. Samuely.

H2   Paper on F.J.S. by Malcolm Higgs. *AA Journal*, June 1960.

H3   Scrapbooks and map of travels in Russia and China. 1933.

H4   Penguin Pool at London Zoo - Draft of reinforcement layout and part of calculation. Photograph during construction, 1933.

H5   Scrapbook of London Period 1933-1943.

H6   Early papers on welding:
'The influence of welding on the design of steel construction'. Welding symposium, 1935.

'The Welding and Cutting Year Book 1936/37'.

'Modern Roof Construction', The Welding Industry, September 1937.

'Welding for Industrial Buildings', The Welding Industry, June and July, 1940.

H7   Records of telephone conservations, 1938.

H8   The Rotunda, Folkestone. Extract of calculations, 1937.

H9   Draft of paper on uses of timber in modern structures, 1937.

H10   Samuely and Hamann, 'Building Design and Construction', Vol 1, 1939.

H11   Manuscript of Vol 2, 1938-39.

H12   Preparation of information sheets (Braithwaite/Architectural Press), 1939.

H13   Palace Gate Flats. Extract from the Architectural Review, 1939.

H14   Samuely and Hamann, 'Civil Protection', The Architectural Press, 1939, and work on air raid precautions, 1938-40.

H15   PATENTS:
No. 3275/39: 31 January, 1939
'Improvements in or relating to preformed ferro concrete structural members, and processes for their manufacture'.

No. 18386/39: 23 June, 1939
'Improvements relating to concrete and the like'.

No. 19458/39: 4 July, 1939
'Protecting buildings against debris load'.

No. 13868/40: 4 September, 1940
'Improvements relating to internal combustion engines'.

H16   Manuscript of the book of welded structures and illustrations, 1940-41.

H17   Draft of booklet on concrete structures for the use of architects for the Cement and Concrete Association, 1941.

Suggestions for Scientific Research on Reinforcement Concrete Structures, 1941.

Report on Education of Architects in the use of Reinforced Concrete, 1942.

H18   Notes on an experimental house at Coventry for E.Neel Esq.

H19   Welded tubular steelwork for Scaffolding Great Britain Ltd. Descriptions, calculations etc., 1930-45.

H20   Korn and Samuely (Mars Group), 'A Master Plan for London', 1942.

48

H21 Draft of paper entitled 'Design for Speed' (The Practice of Design), 1945.

H22 Draft of paper on welded tubular steelwork for Inst. Civil Engineers, 1945.

H23 DSIR - Building Research Station. Testing of trussed beam, 1949.

H24 Reports of tests of structural components.

H25 Lecture notes for the Architectural Association, 1945 onwards.

H26 Paper on Force and Form, *RIBA Journal*, Third Series, Vol. 56 No.2, March 1949.

H27 Paper on Space Frames and Stressed Skin Construction, *RIBA Journal*, Third Series, Vol.59 No.5, March 1952.

H28 Paper on Structural Prestressing, Inst. of Structural Engineers, Vol.XXXIII No.2, February 1955.

H29 Patent on Tensteel Beams No. 23502/57, 24 July, 1957, 'Improvements in Structural Steel Sections', 1957.

H30 De La Warr Pavilion, Bexhill (Mendlesohn & Chermayeff). First all welded steel structure in UK, 1935.

H31 Dovetail Metal Sheeting Factory for Steel Ceilings Ltd., 1936.

H32 Simpsons, Piccadilly (Emberton). Welded steel structure, 1936.

H33 Whittingehame College (Pilichowski). Welded steel frame with dovetail steel decking, 1936.

H34 Sunspan House (Wells Coates), 1926.

H35 House at Esher (Wells Coates), 1936.

H36 Place Gate Flats (Wells Coates), 1937.

H37 News Chronicle Schools competition (Wells Coates & Lasdun), 1937.

H38 House at Frognal (Connell, Ward & Lucas), 1937.

H39 Offices for Gilbey, Camden Town (Chermayeff). Reinforced concrete frame, 1937.

H40 The Rotunda, Folkestone (D. Pleydell-Bouverie). Reinforced concrete dome, 1937.

H41 Laboratories for ICI Manchester (Chermayeff). Reinforced concrete frame, use of reinforced brickwork, 1938.

H42 L.M.S. Garages (Scaffolding Great Britain), 1939-45.

H43 Heston Aerodrome Stores Building (S.G.B), 1939-45.

H44 Juxon St. Police Garages (S.G.B.), 1939-45.

H45 Messrs. Truvox Ltd. (S.G.B.), 1939-45.

H46 S.G.B. Standard Trusses (S.G.B.), 1939-45.

H47 Factory at Malago, Bristol for E.S. & A. Robinson Ltd. (John E. Collins). Precast and prestressed concrete frame, post-tensioned concrete foundations, 1949.

H48 Proposed hangar for B.E.A., 1949.

H49 Saltash Bakery (Bertram Carter). In situ concrete folded plate, 1950.

H50 Kingsmead School (Fairweather). Assembly Hall roof. Composite concrete folded plate, 1950.

H51 Hatfield Technical College (Easton & Robinson). Precast concrete frame, 1950.

H52 Woodberry Down Comp. School (LCC Architects Dept.). Assembly Hall gallery. Concrete latticed folded plate, 1951.

H53 Woodberry Down Comprehensive School (LCC Architects Dept.). Assembly Hall roof. Latticed steel folded plate, 1951.

H54 Power and Production Pavilion, Festival of Britain (Grenfell Baines). Welded tubular steelwork, 1951.

H55 The Skylon, Festival of Britain (Powell & Moya). Pretensioned steel cables, 1951.

H56 Wigan Technical College (Grenfell Baines). Prestressed steel folded plates, 1952.

H57 Wigan Technical College (Grenfell Baines). Assembly Hall roof. Composite concrete folded plate, 1952.

H58 Blackwell School, Harrow (Stillman & Eastwick-Field). Prestressed brickwork, 1952.

H59 Blackwell School, Harrow (Stillman & Eastwick-Field). Covered ways. Composite concrete folded plate, 1952.

H60 Trinity Church Poplar (Handisyde). In situ concrete frame, 1952.

H61 Unitectum prestressed steel roof system (Hartland Thomas), 1952.

H62 Silentbloc Factory, Crawley. Northlight roof. Welded tubular steel folded plate, 1963.

H63 St.Clement Danes Comprehensive School (LCC Architects Dept.). Timber and prestressed timber folded plates, 1953.

H64 School in Beckenham, Kent (Mayorcas). Prestressed concrete trussed beams, 1953.

H65 Ipswich Congregational Church (John Slater & Haward). Precast post-tensioned folded plate roof, 1956.

H66 Cow Close School (John Slater & Haward). Cantilever prestressed concrete roof trusses, 1957.

H67 Stamford Church, Connecticut (Harrison & Ambrowitz). Precast concrete folded plates, 1957.

H68 Ystrad Mynach College (Alwyn Lloyd & Gordon). Prestressed concrete cantilever trusses to floors and roof, 1958.

H69 American Embassy (Saarinen). Precast concrete frame, 1958.

H70 Brussels Exhibition, British Government Pavilion (Howard V. Lobb). Timber folded plate, 1958.

H71 Fisons Office Block, Ipswich (John Slater & Haward). Precast concrete frame incorporating Tensteel Beams, 1959.

# ACKNOWLEDGEMENTS

The Architectural Association is grateful to Frank Newby and Julia Elton for all their work in selecting and assembling the material for this exhibition and their generosity in lending items from their own collections to the AA for the duration of the exhibition. The AA is also grateful to the following for the loan of material:

The Institute of Civil Engineers
Ironbridge Gorge Museum Trust
The Science Museum
Mr Tom Peters of ETH Zurich
Mr Ben Weinreb

The exhibition and catalogue have been organised at the Architectural Association through the office of the Chairman, Alvin Boyarsky, assisted by Micki Hawkes and the Communications Unit co-ordinated by Dennis Crompton with Christine Wallace, Virginia Charlery (catalogue) and June McGowan (exhibition).

Printed in London by Spin Offset Ltd.

The Architectural Association,
34-36 Bedford Square, London WC1 3ES.

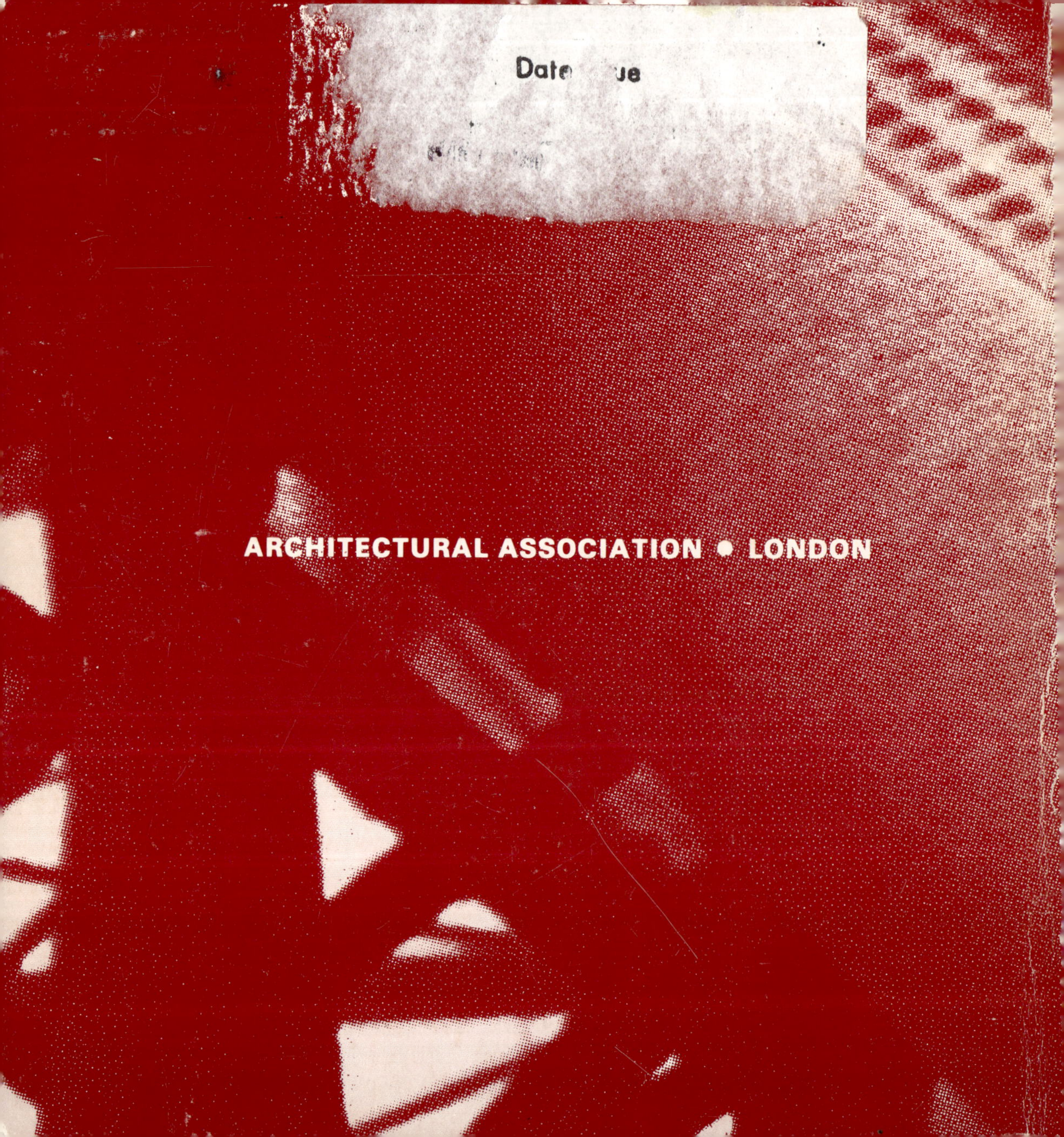

Date Due


ARCHITECTURAL ASSOCIATION • LONDON